Debris Flow

Phenomenology and Rheological Modelling

Debris Flow

Phenomenology and Rheological Modelling

by
G. Lorenzini and N. Mazza
Alma Mater Studiorum - University of Bologna
Italy

WIT*PRESS* Southampton, Boston

Giulio Lorenzini and Natascia Mazza
Alma Mater Studiorum – University of Bologna
Italy

Published by

WIT Press
Ashurst Lodge, Ashurst, Southampton, SO40 7AA, UK
Tel: 44 (0) 238 029 3223; Fax: 44 (0) 238 029 2853
E-Mail: witpress@witpress.com
http://www.witpress.com

For USA, Canada and Mexico

Computational Mechanics Inc
25 Bridge Street, Billerica, MA 01821, USA
Tel: 978 667 5841; Fax: 978 667 7582
E-Mail: infousa@witpress.com
http://www.witpress.com

British Library Cataloguing-in-Publication Data

A Catalogue record for this book is available
from the British Library

ISBN: 1-85312-802-3

Library of Congress Catalog Card Number: 2003109776

Printed in Great Britain by Lightning Source, Milton Keynes.

Cover photo: The propagation of a wave during the debris flow at Tahoma Creek from Mount Rainier, July 26, 1988
(photo credit: U. S. Geological Survey/photo by G.G. Parker).

Table of Contents

Foreword

Due to the increased frequency with which debris-flows occur and the consequential impact they have on human life, in recent years these parossistic events and related matters have attracted increasing attention from research groups and land planning and management professionals. Debris flows are amongst the most frequent and destructive of all geomorphic processes, they mainly affect mountainous areas in a range of morphoclimatic environments and the damage they cause is often devastating. Increased anthropisation calls for improvements in the criteria used to identify debris-flow risk areas and the prevention measures adopted. Phenomenological analysis defines a debris-flow as mass wasting involving granular mass composed of scarcely sorted sediments, in the form of deposits in the mountain portion of basins or in some cases, generated by surface erosion, landslides or volcanic eruptions, that are shaken and mixed with water and set in motion when flooding occurs, thus creating flood waters containing sediments that move downhill with velocities of up to several tens of metres per second. One of the most important aspects of debris-flows is therefore the transformation of solid mass into liquid mass due to the progressive incorporation of water and they are often defined as water "saturated" sediment mixtures, without specifying exactly what is intended by the "saturation" of the mixture. In actual fact, saturation is a particularly delicate concept of soil hydraulics and is directly connected to that of the soil's water potential; it implies that the spaces between grains are entirely full of water and, therefore, although the system undoubtedly has a "high degree of saturation" or is "almost saturated", saturation would not appear to be an essential condition in causing a debris-flow to occur. It therefore seemed apt to include a section that focuses on this particular matter, which can be found in Chapter 3. For the sake of convenience, the terms "saturation" or "saturate" will be used, but the points made in this foreword and those specified later should be considered.

Phenomenological analysis also frequently defines debris flows as gravitational sediment movements, gravity being the dominant driving force. The granular motion regime that characterises this phenomenon is halfway between the quasistatic regime typical of landslides and the hydrodynamic regime typical of diluted suspensions. The most characteristic aspects of debris-flows are the mutual interaction between the solid and fluid components and the inertial effects linked to intergranular collisions and the deformation of the mixture.

The principle characteristics of debris-flow mechanics, which should be considered fundamental in any attempt at modelling, are dilatancy, internal friction, cohesion, fluidification, segregation of the particles and the mechanisms responsible for the dispersion of the granules within the flow. There are four key characteristics to the transport of momentum in this phenomenon: the viscosity of the interstitial fluid, turbulence, sliding friction and the collisions between particles. Since the outset, the aim of debris-flow research has always been the comprehension of the physical mechanisms that govern the triggering of the phenomenon along particularly steep slopes, mobilisation and the relatively stoppage halt in the fan areas. The broadness of the question requires approaches from various perspectives; the dynamic behaviour of these hyperconcentrated water-sediment mixtures and the constitutive laws that govern them, given their non-Newtonian nature, are continuous sources of interest. Debris-flow modelling requires a rheological model (or constitutive equation) that provides an adequate representation of these water-sediment mixtures. Since Bagnold presented his debris-flow model based on the dilatant fluid hypothesis in 1954, numerous other constitutive laws have been proposed for granular material subject to continuous stress. One of the main difficulties encountered by the approaches illustrated in the literature is linked to their possible validation either in the field or in a laboratory environment. On the one hand, the recurrent use of semiempirical relationships referring to average flow parameters, limits the practical applicability to real situations, which are often extremely heterogeneous and on the other hand, by definition, despite their intrinsic generality, the use of exclusively theory-based models is likely to suffer a number of applicational restrictions due to their excessively complex nature caused by them being so far removed from the phenomenon's real context. The choice of a rheological model is therefore one of the most important aims in debris-flow modelling. This book aims to provide methodological details, which can be applied to investigations on debris-flow mechanics that are capable of providing an accurate representation of the phenomenology. The book considers the following subjects.

Chapter 1 provides details on the ambit within which the study was performed and the context within which it lies. Precisely, by making the preliminary consideration that debris flows are an intermediate phenomenon between mass wasting and the processes of solid transport in streams, an initial overview is given on mass wasting and the solid particle transport modes that exist within a flow, according to the forces acting on it.

In Chapter 2, starting from the wide set of hyperconcentrated currents, analysis focuses on the particular case of debris flows and, despite the fact that natural processes can evolve gradually from one another, attempts are made to identify a biunivocal correspondence between terminology and process, by means of a classification criterion based on the rheological behaviour of the material; certain aspects of interest in the rheological study of debris-flows have been included and, by shedding light on the properties of the mixture, they are fundamental for the creation of effective prevention and risk-area forecasting measures.

By means of a study backed up by thorough bibliographic research on field observations and data, laboratory experiments and deposit analysis, in Chapter 3 attempts are made to outline the phenomenological framework of this complex phenomenon.

Chapters 4 and 5 include the presentation of a critical analysis of the principle concepts and theories used in the formulation of the various rheological models provided in the literature, highlighting their strong and weak points and then attempting an overview of the current state of the art in research on debris-flow modelling concerning trigger mechanisms and debris-flow dynamics in steady motion.

In Chapter 6, the question of debris-flow modelling is tackled from a general point of view, starting from the studies of the system's mass balance and momentum and from the various types of stress that act within the flow.

Chapter 7 focuses on presenting a number of plans of action for reducing the risks and hazards of debris flows.

The principle conclusions reached in this book are discussed in Chapter 8, as well as a number of proposals for future studies.

Finally, we would like to express our thanks to Dott. Com. Andrea Brigliadori for his precious advice and to Communication Ways (Bologna, Italy) for the translation.

G. Lorenzini and N. Mazza

Chapter 1

Debris flows: an intermediate phenomenon between mass wasting and solid transport

1.1 Introduction

In recent years, debris-flows have attracted increasing attention from research groups and land-management professionals. Debris flows are among the most frequent and destructive of all geomorphic processes, they affect mountainous areas in a range of morphoclimatic environments and the damage they cause is often devastating. The increase in anthropisation of the mountainous areas of a number of countries worldwide calls for improvements in the criteria used to identify debris-flow risk areas and the prevention measures adopted [1]. The international debris-flow research scene is both complex and diverse, due, at least in part, to the particular characteristics of these phenomena, whose typical peculiarities are a cross between mass-wasting and solid-transport processes in streams.

1.2 Mass wasting

The erosion and subsequent deposition cycle is one of the natural processes that shape the Earth's surface; the varying geographical characteristics that result are a reflection of this continuous dismantling and reconstruction that occurs in nature. Steep mountain slopes are usually characterised by numerous jagged or steep channels or precipices on hillsides, whose slope encounters the gradual declivities of the vegetation covered valleys, before joining the gentler ones of alluvional lands. The latter have always provided a particularly attractive location for human settlements due to their fertile nature and pleasant topography, which make them ideal for both farming and residential development. Moreover, should the world's population continue to increase, these areas could become the only ones available for further urbanisation. Despite the fact that the above-mentioned slopes generally appear stable and imperceptible, they undergo constant alterations due to the gravitational force that causes the movement of the materials that constitute them. This movement maybe imperceptible and gradual, but it can also culminate in catastrophic avalanches or collapses. Unfortunately, the explosive growth of urbanisation puts a strain on the balance of these seemingly inactive environments, with disastrous consequences.

The impact of these phenomena in terms of victims goes hand in hand with the increase in human penetration over the territory. In less developed areas, this loss may be even greater due to population density, a lack of adequate territorial regulations, insufficient public information and the inadequacy of emergency measures. In alpine areas, such phenomena may be exacerbated by the retreating of glaciers and permafrost caused by the radical changes in the world's climate. Better planning and reduced vulnerability and risk from such events may result from the knowledge of the bonds between local geological connotations and fall processes [2].

Mass-wasting phenomenology includes the erosion of entire material accumulations rather than individual particles and all the processes involving the movement of rocks and earth along slopes as a consequence of the effect of atmospheric agents and rock fragmentation.

The phenomenon has three basic factors: the nature of the slope (coherent or noncoherent), the quantity of water in the material, the gradient and instability of the slope; these last two being particularly closely linked to human activity.

1.2.1 The nature of slope materials

Incoherent materials have as a maximum angle of slope, the angle of repose that generally varies between 25° and 45° (approximately 35° for loose sand). The angle of repose is defined as the maximum angle at which a pile of incoherent granules remains stable due to intergranular friction. This angle depends on particle shape and size; wide, flat particles have larger angles, as do scarcely sorted and partially saturated sediments. Slightly damp sand has greater angles of repose than dry sand, due to the film of water surrounding the particles, which develops superficial stresses that hold them together, thus creating negative water pressure between one grain of sand and another (this is the concept of water potential in the various types of soil) and therefore attraction forces along the segment connecting the two grains. Wet sand, however, has smaller angles as in eliminating solid contacts, the water permits the material to run off; which is generally further aided by the increase in the pressure in the interstitial fluid, which reduces resistive strength [3].

Coherent materials do not have an equivalent to the angle of repose and only become instable if excessively wet or devoid of vegetation, they are also resistant to movement as a consequence of internal friction, which is particularly influenced by the water content; they can be divided into cemented materials or conglomerated noncemented materials thanks to cohesion, which acts between different types of particles.

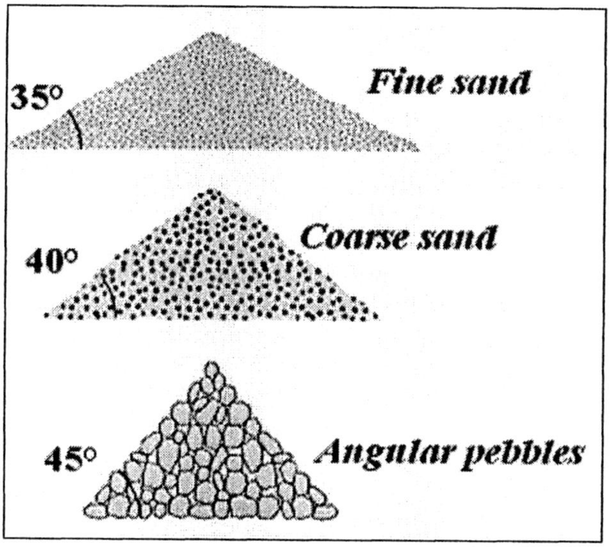

Figure 1.1: Angle of repose for some unconsolidated materials.

1.2.2 Trigger factors

Gravity is the active force responsible for any event linked to mass wasting; it goes without say that the greater the inclination, the greater the tendency of the grains to shift along the slope. Bedrock direction may significantly alter slope stability; when it is inclined in the same direction as the slope, it will be less stable (for example, gravely slopes). This tendency is hindered by the cohesive strength of the material and the friction on the scabrous plane. The effective movement of slope materials is a consequence of its instability; on motion triggering, the particles' potential energy is converted into kinetic energy.

Mass wasting occurs if the gravitational force acting on a material prevails over its resistive force. According to Newton's first law of motion, a body remains in the same state unless an unbalanced force acts upon it. Therefore, in the equilibrium condition, there is the balance between the force of gravity and the (cohesive) forces caused by internal friction. Whereas gravity is more or less constant the world over, the resistance of materials can vary from one time or place to another; the inclination and type of material varies with place, while humidity varies with time. On a flat surface, gravity pushes debris in a direction perpendicular to the surface, whereas along a slope it may be decomposed in the component perpendicular to the inclined plane, which tends to hold materials in place due to friction, and in the parallel component, which tends to move them downhill; as the gradient increases, so the parallel component tends to prevail, which has a destabilising affect.

Water is extremely important to the stability of a slope and has an active action, by increasing the weight of the sediment, or a passive one, by reducing friction and cohesion between the particles, especially if it penetrates between a permeable and an impermeable layer. In this case, due to the polar structure of water molecules, they tend to come between the contacts between polar or ionic solid layers. As mentioned previously, when the debris has a low water content, there is an increase in resistance of the material to movement, however, when the water content is high (closer to a condition of "saturation" of the material) the opposite occurs.

Mass wasting is induced by various factors that trigger movement: violent rainfall, earthquakes (which cause liquefaction and help the force of gravity to prevail over static friction), tectonic activity (for example in the Himalayas and in California), a gradual increase in gradient due to erosion, human activity (artificial damming, civil engineering works that alter natural balance), lowering of the bed following rockslides along slide planes [1].

1.2.3 Introductory definitions

Each perceivable fall movement in a set of materials, however variable, is often named mass wasting. Generally speaking, it can be defined as the detachment and fall of considerably sized rocky masses due to prevalent gravitational force, especially along considerably steep versants, thus causing an accumulation of the same material downhill, usually at the point of maximum inclination. As mass wasting can take place in a range of processes, which may gradually evolve from one type to another, attempts at rationalising phenomenological typologies and transition limits must reflect the nature of the slope material (consolidated or non consolidated material), the velocity and the nature of the movement that characterise the phenomenon. On the basis of such classification criteria, it is possible to identify two large categories of mass wasting: landslides and sediment flows.

1.2.3.1 Slope failure

This causes the transport of sediment downhill by means of rolling, sliding, falling and land collapsing.

Falls occur when rocks on very steep slopes or overhanging rocks become unstable and fall down the slope; this therefore creates a sudden detachment, usually along a pre-existing discontinuity. During the fall they may interfere with other rocks and cause them to fall in turn. They involve sediment and rocks (and in some cases, as in debris falls, debris) that may either move up into the air or freefall along the ground in a downhill direction. They can occur in the absence of water, triggered by the action of plant roots, or can be damp processes linked to the action of ice. Falls are extremely rapid phenomena (m/min, m/s) and result in an accumulation of material at the foot of the slope, known as the *talus*.

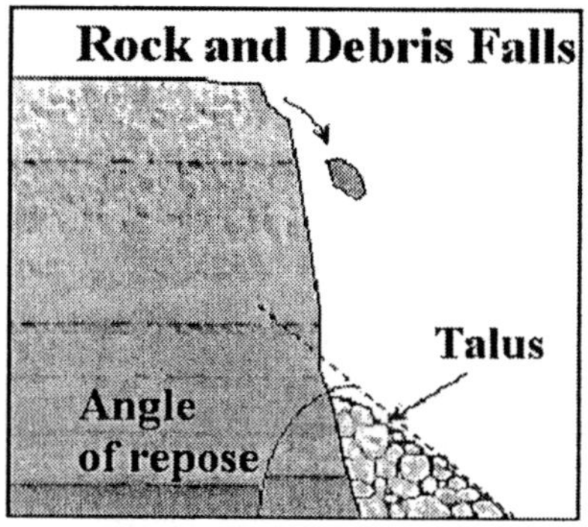

Figure 1.2: Falls involve sediment and rocks that may either move up into the air or freefall along the ground in a downhill direction.

Slides are caused by the descent of rock sheets on several plastic levels and, in some cases, following water infiltration; they therefore represent a type of mass wasting that causes a unitary sliding movement, mainly along pre–existing planar surfaces, such as joint planes (the joints consist in regularly spaced fractures in the rock caused by expansion during cooling or rising); slides differ from *slumps* in that there is no rolling along a curved surface.

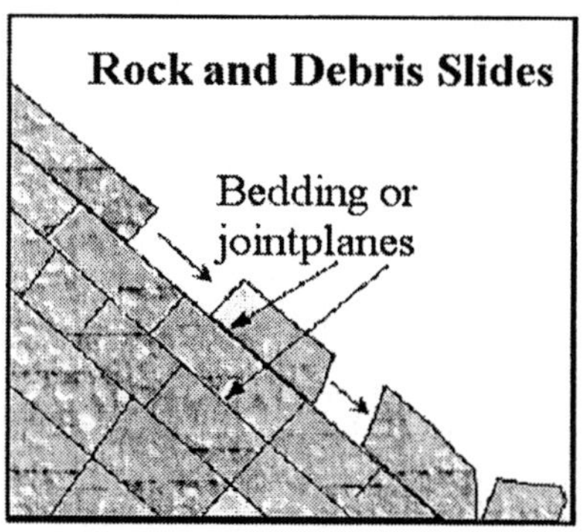

Figure 1.3: Slides represent a type of mass wasting that causes a unitary sliding movement, mainly along pre-existing planar surfaces.

Slumps consist in the sudden and rapid precipitation of rocks such as clay and marl with a consequent downwards rolling (mm/y, m/s) of the material involved along a curved surface with an upwards concavity; the upper surface of each slump remains relatively undisturbed. This kind of event causes depressions in the slope and they often take place as a result of human activity, following excessive removal of material from the foot of the slope, such as that required for road construction or for gravel quarrying, as well as erosion of river banks or seashores; moreover, they may be triggered by heavy rainfall or earthquakes.

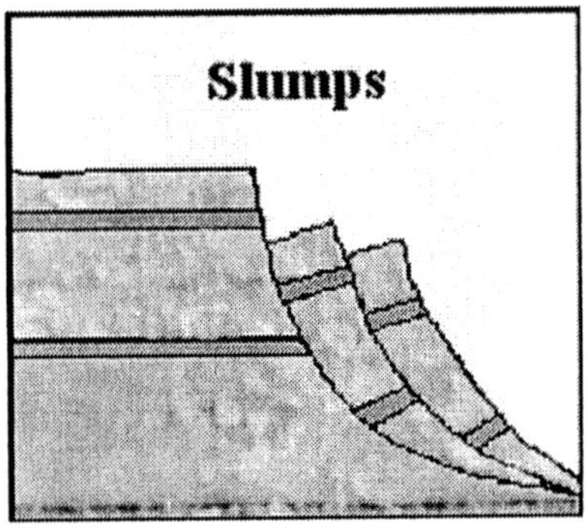

Figure 1.4: Slumps involve a rapid downward rolling of the rocks along a curved surface with an upward concavity.

1.2.3.2 Sediment flows

These occur when a sufficient force is applied to a pile of rocks and earth, thus causing runoff along the slope. They consist in plastic or semi-liquid rock and sediment movements either in the air or on the ground. They include a wide range of mass-wasting types, from the slowest to the most violent and rapid kind and can be divided into two categories according to water content.

1.2.3.2.1 Slurry flows. These are rapid descents of incoherent material, or material that has become incoherent due to water imbibition. Such flows contain between 20% and 40% volume of water, beyond this limit one talks of a stream. Slurry flows are considered flows with a high degree of saturation [1].

Solifluction consists in the slow and continuous movement of "saturated" soil and results in the formation of distinctive lobes on the slopes. Velocities of a few centimetres per year can be reached and this phenomenon is typical of Nordic regions where freezing and thawing alternate cyclically.

Debris flows are mixtures of concentrated air-"saturated" or water-"saturated" sediments, which proceed extremely rapidly along slopes under the action of gravity with velocities of between 1 m/y and 100 m/h; one salient characteristic of debris-flows is the wide range of size of the sediment it contains. The phenomenon occurs following heavy rains when the material becomes "saturated" with water; however, they also frequently arise from slumps or landslides that are transformed into flows during their descent, when the mass disaggregates and mixes with air and water. During the flow they tend to increase in volume and the deposition process results in the formation of lobes with irregular surfaces. They can cover large expanses on fairly regular slopes, thanks to the runoff that occurs by exploiting an air cushion (m/min, m/s).

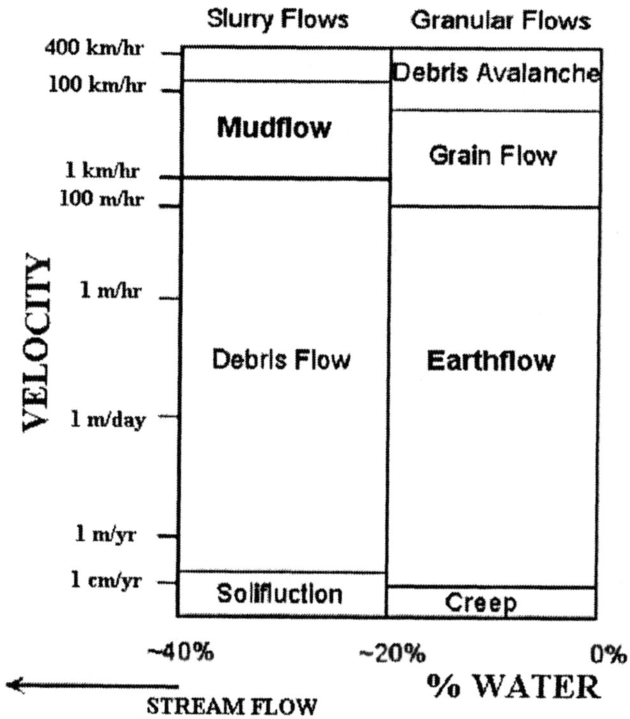

Figure 1.5: Sediment-flow classification according to velocity and water content.

Mudflows have a high water content and a large proportion of fine and cohesive granular sediment (at least 50% is composed of sand, silt and clay) and are therefore extremely fluid. They differ from debris flows in that there is less variation in grain-size. Mudflows can reach maximum speeds of 1 km/h, and can cover great distances, even in valleys, which makes them potentially very dangerous.

1.2.3.2.2 Granular flows These are flows that contain between 0% and 20% volume of water and their semi-fluid behaviour derives from mixing with air; granular flows have a low degree of saturation.

Creep is a continuous, extremely slow movement (cm/y, mm/y) and is the most significant phenomenon in terms of the total volume of material transported per year. This phenomenon is also facilitated by the cyclical cooling and warming of the soil.

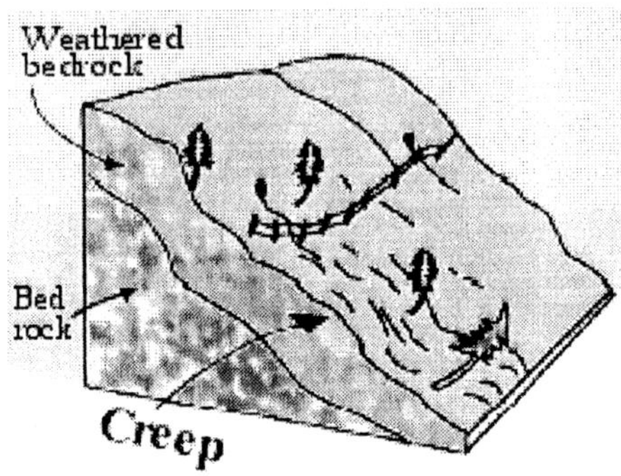

Figure 1.6: Creep is a continuous, extremely slow movement facilitated by the cyclical cooling and warming of the soil.

Debris avalanches are an extreme form of debris-flows, they move at extremely high speeds and shift large volumes of material, deriving from the collapse of entire mountainsides; they resemble snowslips and move along the slope like elongated tongues.

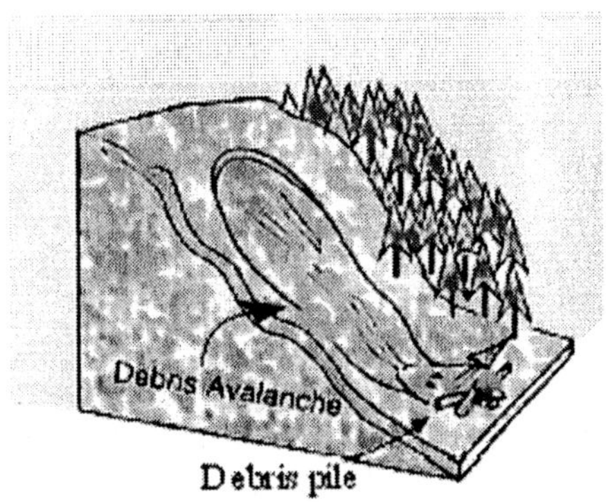

Figure 1.7: Debris avalanches move at extremely high speeds and shift large volumes of material.

Debris avalanches usually initiate from open hillsides or from shallow hillside depressions where phreatic water gathers, they can transform into debris-flows if sufficient water is present.

Table 1.1: Synthetic overview of types of mass wasting grouped according to the characteristics of material, the type of movement and velocity.

		VELOCITY		
MATERIAL	NATURE OF MOVEMENT	SLOW	MODERATE	FAST
CONSOLIDATED	FLOW			Rock avalanche
	SLIDE		 Rockslide	 Rock fall
UNCONSOLIDATED	FLOW	 Creep	Earthflow/ Mudflow	Debris avalanche
		Solifluction	Debris flow	
	SLIDE	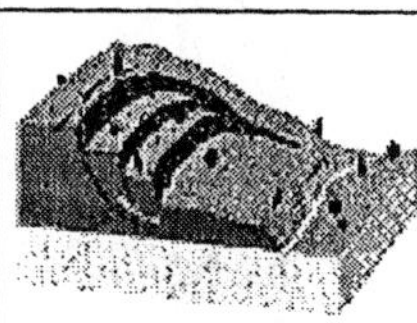 Slump	 Debris slide	

1.3 Solid transport within a streamflow

Having said that debris flows are an intermediate phenomenon between mass wasting and forms of solid transport in streams, it is essential to understand which possibile mechanisms cause the transport of solid particles within a streamflow. Generally, having a greater specific weight than the fluid phase, solid particles tend to deposit and therefore in order to keep them in movement within the streamflow, forces that contrast the submerged weight of the granules and prevent them depositing are required. In particular, being a solid–liquid streamflow, the balance must include the interactions due to intergranular contact between particles (rubbing and collisional contact) and also turbulent interactions in the fluid. According to the type of forces acting, solid particle transport within a flow can be classified as bed-load, suspended load and neutrally buoyant load [4].

1.3.1 Bed-load

In this kind of transport, the submerged weight of the particles, which is weight minus the Archimedes' buoyancy, is counterbalanced by the dispersive forces or contact forces; if dispersive forces prevail, the particles hop and bounce along the channel, if, however, contact forces prevail, they slide and roll along it [4]. In general, the width of the layer within which the transport takes place is approximately the same as the diameter of the particles, therefore the layer of the streamflow in which the phenomenon is localised is, of course, the bottom of the channel. Ordinary bed-load transport occurs when the submerged weight of the granules is sustained by the dispersive stress or the static stresses that arise in contacts. When the average stress around a generic particle on the channel bed exceeds a threshold value, it starts to move on the others in the direction of the flow. The characteristics of movement depend on the conditions of the flow, the density of the two phases and the dimensions of the particles [5].

The phenomenon of bed-load transport can be categorised according to a number of different approaches [6], which are not necessarily mutually exclusive, but rather are somewhat complementary in describing a complex phenomenon.

1.3.1.1 Boundary layer of solids in movement

Dragged by shear stress, a variable layer of solid particles is set in motion. As depth increases, the solid layers are characterised by gradually decreasing velocity.

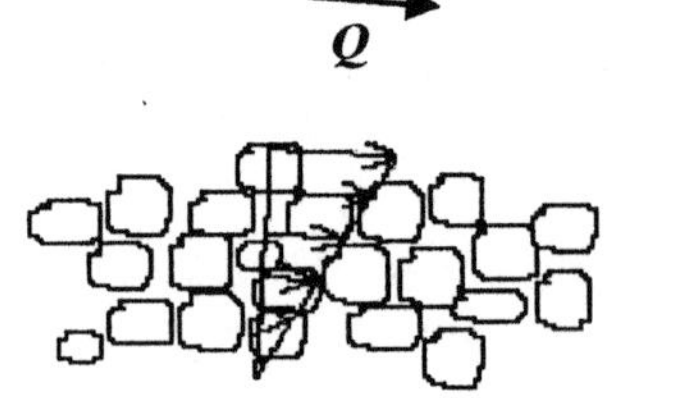

Figure 1.8: Boundary layer of solids in movement.

1.3.1.2 Surface movement

Only a surface layer of solids on the bed is considered to be in movement; the particles of this superficial layer roll/rub on those below, which remain practically immobile.

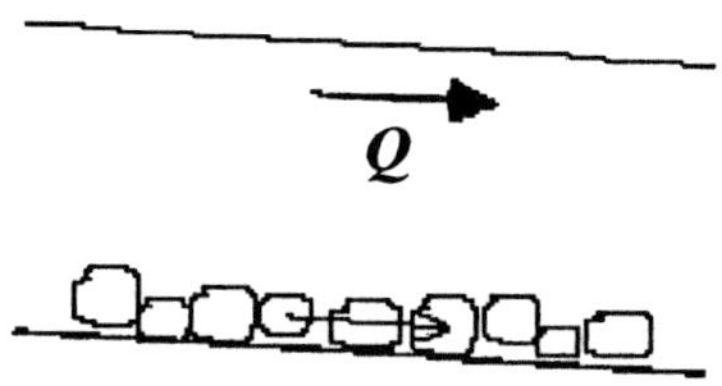

Figure 1.9: Surface movement.

1.3.1.3 Jumps

The movement of the particles is not classified as a continuous phenomenon, rather a series of finished "jumps" of superficial particles, followed by rests.

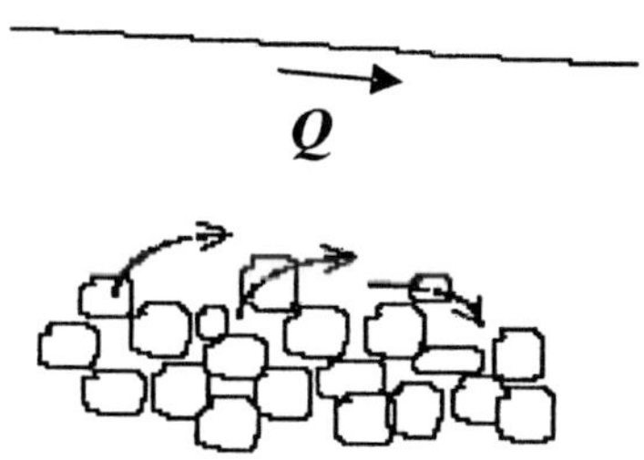

Figure 1.10: Jumps.

In the first two cases, contact load takes place with a prevalence of contact forces, whilst in the third case we talk about saltation load with a prevalence of dispersive forces; this mode is important when noncohesive materials are present with relatively high sedimentation velocity, such as sand in air and gravel in water and can cover the entire depth of the streamload, such as in debris flows with low clay material contents [4]. One bed-load transport mechanics analysis scheme is Bagnold's dispersive model [7] which, given its centrality in the understanding of debris-flows, will be dealt with later.

1.3.2 Suspended load

Due to turbulent components in streamflows, it is possible that solid particles are lifted off the bed and transported, even for considerable distances, held in suspension within the streamflow. The submerged weight of the grains is counterbalanced by the diffusive force that depends on the intensity of the turbulence and the sediment concentration gradient [4]. This phenomenon is, in itself, of considerable importance for certain waterways (such as the Nile and Ganges); another characteristic example is that of dust carried by the wind. This type of transport, which is only possible for small sized sediment (silt, clay and sand) with a low sedimentation velocity, differs from the previous one in that once the particles are in suspension they assume the translation velocity of the streamflow, whilst the particles in movement on the bed move with an average velocity lower than that of the fluid above [8]. As it requires turbulent energy alone, suspended load does not cause greater resistance to flow movement, which instead does take place with bed–load. The amount of sediment transported in suspension is linked to the streamflow's ability to lift the particles and the time the particles are kept in suspension before falling back to the bed (and therefore to the solid concentration within the flow). According to Fick's law, the concentration gradient within a turbulent flow must be accompanied by a diffusive movement and the amount of sediment transported by diffusion through unit area is given by:

$$q_s = \varepsilon_s \nabla(C_v), \tag{1.1}$$

where ε_s is the sediment diffusion coefficient and $\nabla(C_v)$ the suspended sediment concentration gradient (see section 2.2); the coefficient is only constant in a homogeneous turbulent flow [4]. In general, it is assumed that ε_s is proportional to the momentum diffusion coefficient ε_m

$$\varepsilon_s = \beta \varepsilon_m . \tag{1.2}$$

Generally speaking, it is difficult to quantify the value of coefficient β and calculate its variation with the fluid's turbulence characteristics.

1.3.3 Neutrally buoyant load

In a hyperconcentrated streamflow (see section 2.2), the presence of clay and silt can affect the rheology of the water–sediment mixture, thus causing a non-Newtonian behaviour with yield stress. Part of the sediments can therefore be sustained by the yield stress that prevents its being deposited; this portion is generally indicated as neutrally buoyant load [4], and is characterised by the absence of relative movement between fluid and grains, therefore, if within the streamflow there are no other transport mechanisms, we can talk of a pseudo monophase fluid. If a streamflow transports enough clay material to alter its rheological characteristics, the main transport takes place by floating and the flow can be treated as a monophase fluid; in this case, the transport flow competence has no sense, as flow characteristics depend on the rheology of the mixture and, in particular, on the cohesive fine fraction content [6]. Transport takes place by high energy waste caused by the system's viscosity. It is possible to estimate the maximum diameter D_0 of the particles that can be sustained by yield stress τ_y [4]:

$$D_0 = K \frac{6\tau_y}{\gamma_s - \gamma_f}, \tag{1.3}$$

in which γ_s is the specific weight of sediment, γ_f is that of the interstitial fluid and K is an experimental constant near to unity. In a non-Newtonian hyperconcentrated flow the portion of sediment with a diameter smaller than D_0 therefore constitutes the neutrally buoyant load [4].

1.4 Observations

The main factors that govern the solid transport mechanisms within a hyperconcentrated flow are the composition, the grain-size and physical characteristics of the sediment (specific weight, shape, angle of internal friction and cohesion), as well as hydraulic factors such as velocity, turbulence intensity and slope angle [5]. As a function of these determining factors it is possible that inside a given streamflow, more than one type of transport may occur simultaneously, thus resulting in intermediate solid transport modes, deriving from the combination of several mechanisms. In general, it can be said [4] that at high concentrations a non-Newtonian behaviour with yield stress prevails and therefore the finer fraction is transported by a neutrally buoyant load and the coarser fraction by suspended-load or bed-load transport. It is, however, possible that within the entire thickness of the flow, there are zones characterised by a predominance of a particular type of transport. A thorough analysis is therefore required of the stresses acting within the solid–liquid mixture, which can provide valid elements for distinguishing the behaviour of the mixture.

Chapter 2

The rheology of hyperconcentrated flows

2.1 Introduction

The most distinctive characteristics of debris flows are the interaction between the solid and fluid components and the associated forces. If, on the one hand, this makes it possible to distinguish debris-flows from other correlated phenomena in which just one of the components is dominant (for example, in landslides and streamflows), on the other, it highlights a vast range of processes that could be included in the same category, such as debris slides, debris torrents, debris floods, mudflows as all these phenomena, debris flows included, belong to the same more general class of hyperconcentrated sediment mixtures [9]. As indicated in the opening overview on mass wasting, in nature, such phenomena take place within a continuum of these conceptual classifications and it is often difficult to distinguish the nature of a given process, partly because it is usually characterised by distinct phases and it is only by fully understanding the phenomenon's intrinsic mechanical mechanisms that one can provide elements to support this hyperconcentrated flow rationalisation process. Studies of such phenomena have given rise to a number of different terminologies, most of which are based on the qualitative differences in the style and characteristics of movement. The evolution of models and the availability of increasingly accurate data call for reference to classification criteria based on the rheological behaviour of the material alone [10].

Therefore, in this chapter, analysis starts from an examination of the set of hyperconcentrated flows as a whole, to focus later on the case of debris flows in particular, despite the fact that, in nature, these processes evolve from one to the other; attempts are made to identify a biunivocal correspondence between terminology and process, using rheologically based classification criteria. This is followed by the examination of certain interesting aspects of debris-flow rheology, which are essential in engineering effective debris-flow prevention systems and risk-area forecasting measures.

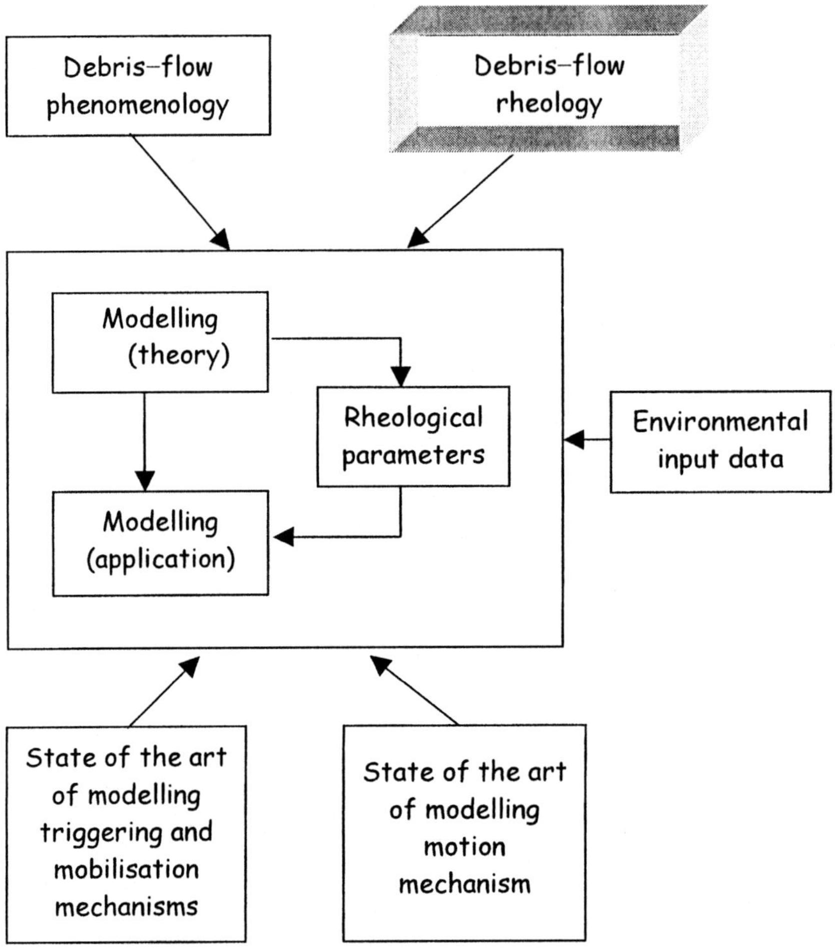

Figure 2.1: Chart showing the debris-flow study approach adopted in this book. In particular, the grey frame highlights the field of analysis concerning debris-flow rheology, which will be the object of studies in this chapter.

2.2 Hyperconcentrated flow classification

A hyperconcentrated streamflow is defined as a fluid in movement in which a high percentage of solid material is transported. Given a liquid–solid two-phase mixture in static conditions, the mean solid concentration by volume is defined as the ratio between the volume occupied by the solid fraction and the total mixture volume: $C_v = \frac{V_{solid}}{V_{solid} + V_{liquid}}$, in which V_{solid} and V_{liquid} are the volume of the solid fraction and that of the liquid fraction of the mixture, respectively. Hereinafter, reference will be made to this definition when discussing concentration by volume or volumetric concentration.

Despite the implicit reference to a two-phase fluid composed of solid material and water, in a hyperconcentrated streamflow the two components do not always remain completely separate and distinct from one another, at least from the point of view of their mutual interaction. In fact, in certain cases we must refer to an effectively two-phase mixture and in others to a homogeneous solid–liquid streamflow.

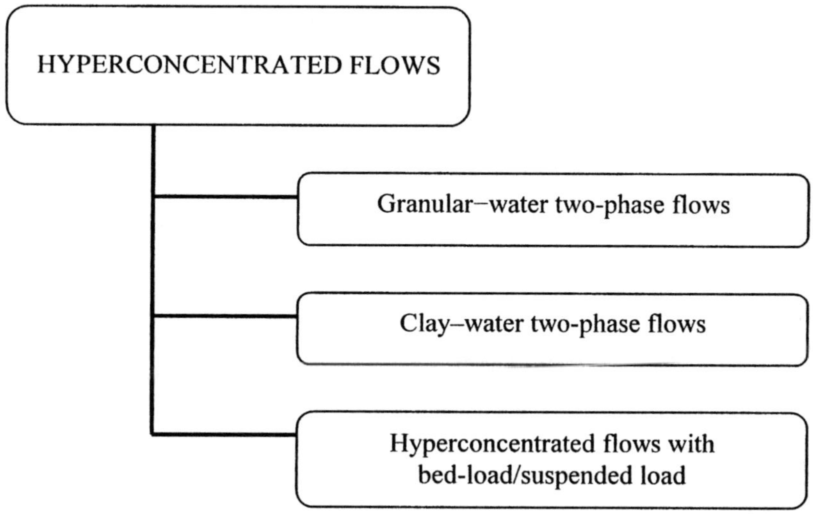

Figure 2.2: Characterisation of the context of hyperconcentrated streamflow studies.

The chart above, which is a natural extension of the definition of hyperconcentrated flows given previously, shows a somewhat extensive set of phenomena, however, each one has its own particular characteristics. From a qualitative point of view, it can be said that water and solid particles sometimes behave as a homogeneous fluid, and sometimes as two separate phases; a distinction that certainly represents the point of contact between the various classifications of two-phase mixtures that have become widespread in studies on hyperconcentrated sediment mixtures.

2.2.1 Main classification criteria

In the past, the main classifications of water–sediment mixtures were based on criteria obtained from direct observations, experimental process evaluations, morphological analysis of deposits, physical models or theoretical studies (Table 2.1).

Table 2.1: Main hyperconcentrated flow classifications used in the literature.

AUTHOR	*CRITERIA*	*CLASSIFICATION*
Sharpe (1938) [11]	▪ relative velocity ▪ relative sediment concentration	▪ *debris avalanche* ▪ *mudflow* ▪ *earthflow* ▪ *solifluction* ▪ *creep* ▪ *streamflow*
Varnes (1958) [12]	▪ type of material ▪ type of movement	▪ *block stream* ▪ *debris avalanche* ▪ *debris flow/mudflow* ▪ *earthflow* ▪ *solifluction* ▪ *creep*
Takahashi (1981) [14]	▪ grain–support mechanisms	▪ *debris flow* ▪ *pyroclastic flow* ▪ *sturzstrom* ▪ *fall*
O'Brien and Julien (1985) [15]	▪ sediment concentration ▪ flow properties	▪ *mudflood* ▪ *mudflow* ▪ *debris flow*
Coussot (1992) [16]	▪ relative amount of fine fraction	▪ *muddy debris flow* ▪ *granular debris flow*

The first mass-wasting classifications that were introduced were concentrated in particular on landslides, a phenomenon that is of great interest to any new urban settlement. Sharpe [11] considers two main parameters, relative velocity and sediment concentration and despite the fact that he does not specify transition boundaries, this classification has been widely used and refers to the following process categories: *debris avalanches, mudflows, earthflows, solifluction, creep* and *streamflows*. Along the same lines as Sharpe, Varnes later presented his classification [12] that became a reference point for the terminology of these processes. Varnes' classification is based on two main characteristics: the type of material and the type of movement involved, whereas velocity and mixture composition are used for subclassification purposes. Therefore, for coarser materials it identifies the following phenomena: block streams, debris avalanches, debris-flows (mudflows if the coarser material content is lower than 50%), solifluction and creep; and for finer materials: blends of dry sands and silt, blends of wet sands or silt and earthflows. In the Chinese literature [13], the term 'hyperconcentrated' is generally used to indicate a material that has a measurable yield stress and therefore debris-flows are considered as hyperconcentrated flows. However, more recent classifications have attempted to systemise established terminology by introducing quantitative criteria.

Takahashi [14] defines mass wasting as the fall, slide or flow of a conglomerate or dispersed mixture of sediment in which gravity moves all the particles and the interstitial fluid, so that the relative velocity between the solid and fluid phases in the main direction of motion, merely plays a minor role, whereas in a fluid flow, the forces of lift and resistance caused by relative velocity are essential for the transport of each single particle. In this approach, the following four phenomena can be distinguished on the basis of the mechanism that supports the clasts, the properties of the interstitial liquid, velocity and distance reached. *Falls,* in which the single particles move separately with relatively small internal deformation; *sturzstroms*, particularly rapid and destructive events; *pyroclastic flows* that are rapid and explosive events with volcanic origins, in which the suspension mechanism is linked to the expansion of the gas trapped within the flow and, finally, *debris flows* in which the grains are dispersed in a water–clay interstitial fluid. These last three processes can be termed collectively as *gravitational sediment flows* [14] and constitute continuous processes that require a certain force for grain suspension. With debris-flows, one approach [7, 14] considers the dispersive pressure that results from the exchange of momentum between grains as predominant; if the interstitial fluid is particularly dense, large clasts can be suspended with relatively low dispersive pressures by floating in the fluid phase. Another approach [9] considers the viscous stress of the interstitial fluid predominant and neglects interactions between grains.

Following various laboratory experiments on samples from Colorado, O'Brien and Julien [15] classified hyperconcentrated flows, according to the properties controlled by sediment concentrations, as *water floods, mudfloods, mudflows* and *landslides* (Figure 2.3). The characteristic stresses of such processes are yield stress, viscous stress, turbulent stress in the fluid and the dispersive stress caused by the inertial impact of the coarser sediments. Which of these stresses dominates depends on the volumetric concentration of sediment and the percentage of the fine fraction. Despite the fact that the transition between the types of flow is difficult to perceive, according to O'Brien and Julien's approach [15], they can be divided into three categories, which lie between conventional stream flooding on the one hand and landslides on the other. *Mudfloods* are hyperconcentrated flows of cohesionless particles (mainly sand) with limited quantities of cohesive particles, which show characteristics that are typical of fluids, with sediment concentrations by volume of 20–45%. From a hydrodynamic point of view, mudflows have characteristics that are typical of a conventional turbulent flow and resistance to motion depends on the roughness of the channel in which runoff takes place.

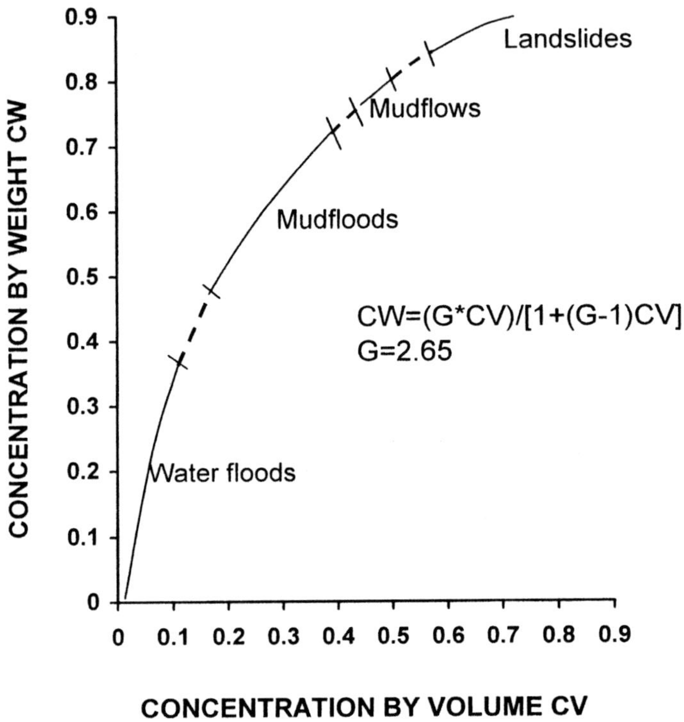

Figure 2.3: The classification of hyperconcentrated streamflows in the study of O'Brien and Julien [15].

Moreover, they are not able to support stress without deforming and show no yield stress. Sediment concentration tends to be uniformly distributed throughout flow depth, because the viscosity of the interstitial fluid reduces the velocity of particle sedimentation.

Mudflows are hyperconcentrated streamflows composed, to a large extent, by cohesive silt and clay particles (smaller than 0.0625 mm) and in which sediment concentration by volume varies between 45% and 55%. This composition alters the properties of the interstitial fluid, making it extremely viscous and giving it considerable yield stress. Consequentially, in the free surface area, mudflows can hold considerably sized clasts in suspension for long distances even on slight slopes, resulting in the formation of lobe-shaped deposits. Typical resistance to motion is a characteristic of pseudoplastic fluids that appears with high viscosity.

Debris flows are mixtures of clastic material with high coarse particle contents, in which collisions between particles and therefore dispersive stresses are the dominant mechanisms in energy dissipation. Cohesionless granular debris-flows are a subcategory of debris-flows, in which momentum exchange takes place due to friction and collisions. These phenomena occur depending on the simultaneous occurrence of the following conditions:

- high volumetric concentration of sediments (>0.5)
- high shear rates (>100 s^{-1})
- large particle dimensions (>5% compared to streamflow depth).

Coussot [16] considers two types of debris flow: *granular* ones, which have a fine particle fraction quantity (dimension smaller than 40 μm) 10% lower than the entire solid mass, and *muddy* ones in which the fine fraction exceeds 10%. The classification proposed is given in terms of dimensional grain distribution, as shown in Figure 2.4.

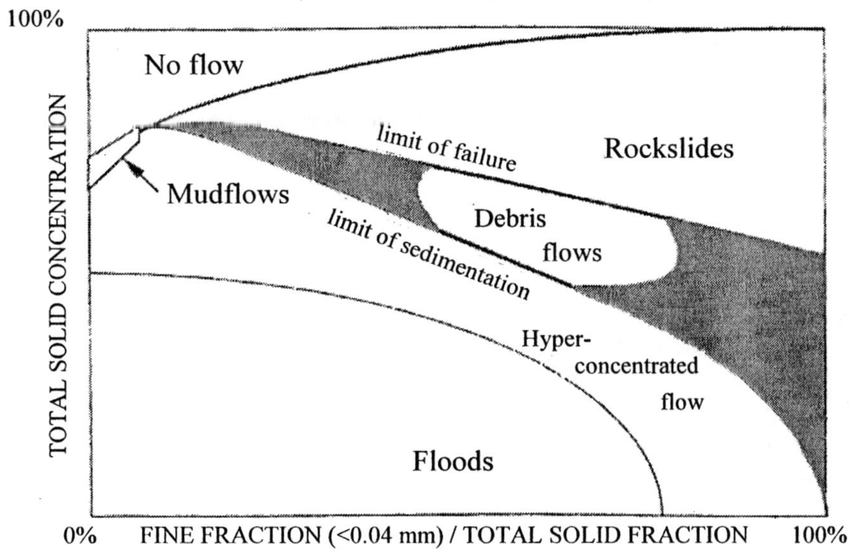

Figure 2.4: Coussot's classification of hyperconcentrated streamflows [16]

In attempts to classify the rheological behaviour of granular mixtures, two further studies must be considered. Savage [17] identifies three flow regimes for granular mixtures, each one characterised by a value of the solid fraction, interstitial viscosity and deformation rate: the *macroviscous* regime, in which the viscous effects of the interstitial fluid and solid particle interactions cause the stresses; the *quasistatic* regime, in which dry friction and prolonged contacts between particles are important, whilst inertial effects are negligible; and lastly, the *inertial-granular* regime, in which the inertia associated to the individual particles prevails. Iverson [18] developed a constitutive equation for the idealised behaviour of mass wasting, based on linear and nonlinear rheological models, which range from the purely plastic case to the purely viscous one; this equation represents an important analytical instrument for differentiating between the various types of flow.

2.2.2 Pierson and Costa's rheological classification

A classification of the various types of flow can be introduced using a bidimensional graph (Figure 2.5) that considers the mean flow velocity and sediment concentration; this classification is proposed by Pierson and Costa [10] and currently constitutes a reference point in this field, as it proposes a univocal classification of hyperconcentrated flows, thus overcoming the terminological confusion that has increased as studies have progressed, as mentioned in the previous overview. This classification makes it possible to distinguish each process from the others, if estimated velocity is known and additional information is available on the existence of yield stress and stream capacity to suspend large clasts, characteristics that can be determined by deposit analysis. The reference material is a mixture of scarcely sorted, cohesionless coarse particles. In the graph, vertical rheological divisions A, B and C depend on grain-size and concentration. From left to right, limit A represents the appearance of yield stress, limit B its sudden and rapid increase that enables the static suspension of granules and the start of fluidification and lastly C, the stoppage of fluidification. The horizontal velocity limits, which are also functions of the previous quantities and particle density, are identified by the stress mechanism that acts between the particles during the flow. In the case of cohesive materials or those that contain a high proportion of fine materials, the vertical lines on the graph, that divide the various rheological behaviour types, must be shifted to the left; the opposite is true if the mixture contains mainly well-sorted, coarse clasts. It is therefore possible to identify two large-flow categories: apparently fluid mixtures and plastic–fluid mixtures. The former includes both *streamflows* and *hyperconcentrated streamflows*.

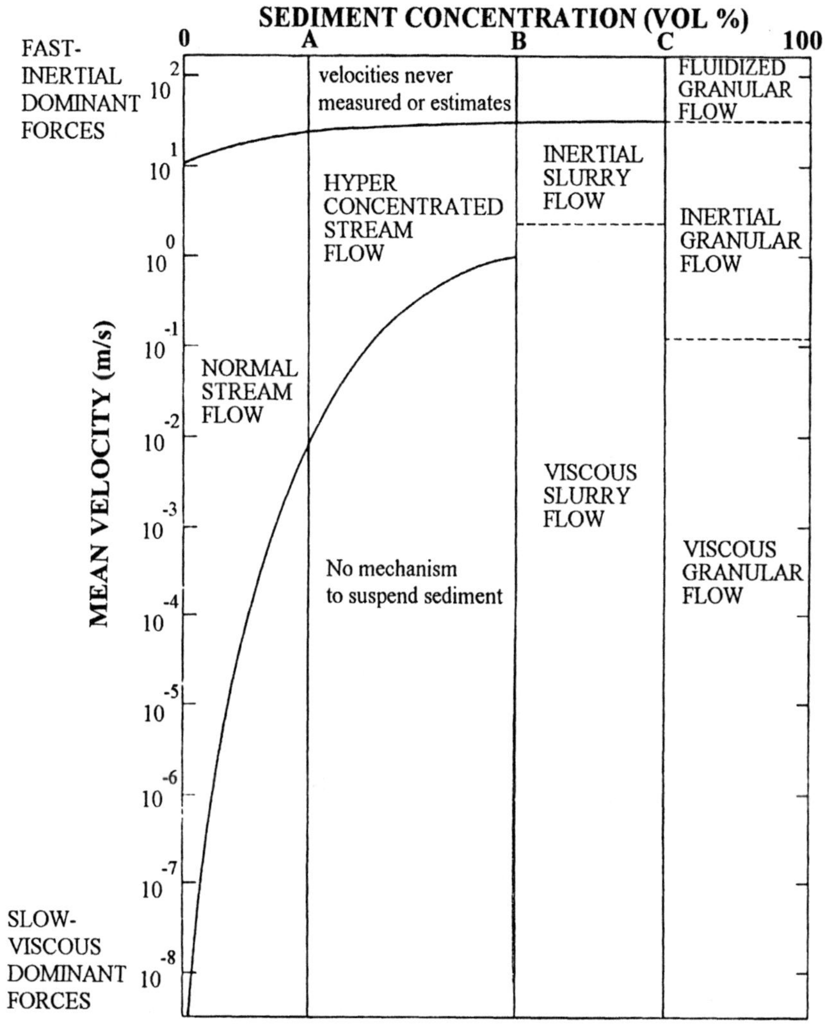

	0–A	A–B	B–C	C–100
FLUID TYPE	NEWTON	NON NEWTONIAN		
PORE FLUID	WATER	WATER+FINES		WATER+ FINES+AIR
CATEGORY	STREAM		SLURRY	GRANULAR
BEHAVIOUR	LIQUID	PLASTIC		

Figure 2.5: Rheological classification of water–sediment mixtures proposed by Pierson and Costa [10].

2.2.2.1 Apparently fluid mixtures

In *streamflows* water is the continuous phase, in which air bubbles or fine organic or non organic sediment particles are dispersed, the suspension is very dilute and therefore there is little interaction between particles and rheological behaviour maintains the characteristics of the continuous phase, in other words, those of a Newtonian fluid [10]. When the concentration of suspended particles increases, the interactions are no longer negligible; if clay particles are also present, electrochemical forces intervene and give the fluid greater resistance, thus giving it a yield stress and therefore making it become non-Newtonian and exceeding vertical limit A in Figure 2.5; this depends largely on grain-size and the presence of clay. In the absence of fine particles, a suspension with low shear rates can remain Newtonian until interactions caused by friction start to prevail (sediment volumetric concentrations greater than 50%).

In the definition given by Beverage and Culbertson [19], *hyperconcentrated streamflows* have sediment concentrations by volume of 20–60% and have a measurable yield stress, but, as this yield stress is rather low, they still flow like liquids.

2.2.2.2 Plastic fluid mixtures

The second flow category, that of plastic mixtures, includes *slurry flows* and *granular flows*.

Slurry flows exceed vertical boundary line B (Figure 2.5), when the increase in sediment concentration is accompanied by a sudden increase in yield stress, which is greater for coarser mixtures, mainly due to sliding friction between grains. In particular, this type of flow has a sufficiently high yield stress to show plastic behaviour (lobe-shaped fronts, lateral banks and coarse grain suspension), but which reverts to being liquefied when stirred [10]. The mixture is "saturated" with water and interstitial fluid pressure is higher than the hydrostatic value (see section 3.7), in such conditions, a part of the weight of the solid phase can be sustained by the fluid itself; once the yield stress is exceeded, the mixture flows as a homogeneous and coherent mass. During stoppage, fine and coarse particles are deposited together without mutual movements. This causes a distinction with those hyperconcentrated streamflows in which particles are deposited separately according to their sedimentation velocity [4]. In viscous slurries, the viscous forces control the behaviour when the amount of clay is high or the mean diameter, granule density and water content are low. For this reason, Bingham's model is often chosen as a constitutive equation. However, if the viscosity of the interstitial fluid is low and shear rate, mean diameter, density and water content are high, inertial forces dominate, therefore momentum is transferred by means of collisions. This gives an inertial slurry flow and often the dilatant model is chosen. The transition of a flow from viscous to inertial can be described by the Bagnold number [7], which is the relationship between inertial and viscous stresses. Due to yield stress, a rigid area could form during the flow, in which the material behaves as a solid, because the yield stress is not exceeded; the more the area dominated by stresses is restricted towards the bottom, the more forms of shearing take place.

In *granular flows* vertical limit C (Figure 2.5) has been exceeded and the sediment concentration is so high that the mass no longer remains partially liquefied when the mixture is stirred [10]. Interstitial fluid pressure is no longer greater than the hydrostatic pressure and any excess pressure is immediately dissipated because the water can run away freely and the entire weight of the granular mass is sustained by contacts or collisions, moreover, as granule concentration by volume is high, direct interactions between them are frequent. *Frictional granular flows* occur at low shear rates and are described by Savage's macro viscous regime [17], in this regime, quasistatic friction and the viscosity of the interstitial fluid dominate. When other parameters are equal, this occurs with mean velocities lower than 0.1 m/s due to the lower viscosity of the fluid phase compared to slurry flows. For this type of process, pure flows are considered an exception, the drop in water content causes the formation of cracks or stress areas along the margins and sliding therefore becomes an important motion mechanism [10]. When sufficient stress is applied to a granular mass to dominate friction resistance, the mass will start to flow, thus involving several layers of particles in relative motion to one another. The slow deformation that takes place constitutes the quasistatic regime. At higher shear rates, the inertial component starts to dominate and momentum is exchanged by means of collisions between grains, thus causing dispersive stress, in the same way as with Savage's inertial regime [17]; mean velocities range from 0.1 to 35 m/s when other parameters are equal. Rapid inertial granular flows occur at high shear rates and high mean velocities that cause the dominance of the energic collisions on friction (debris avalanches and sturzstroms).

2.2.2.3 Comparison between terminology used in practice and Pierson and Costa's rheological classification

The terminology employed in Pierson and Costa's classification [10] is based on the forces that dominate each process (inertial, viscous and friction) and on the presence or absence of a yield stress. It is therefore an attempt at classifying hyperconcentrated streamflows on the basis of rheological behaviour alone and thanks to this, it is also possible to use the limits proposed to conveniently group the terminology that is widely used and was discussed previously (Figure 2.6). Therefore, in accordance with the definitions given in the opening overview, mudflows belong to the hyperconcentrated streamflow category, whilst *viscous* and inertial slurry flows include the wide phenomenological range of debris flows, whereas mudflows are well described by the characteristics of a viscous slurry; and granular flows include sturzstroms, debris avalanches, earthflows and creep.

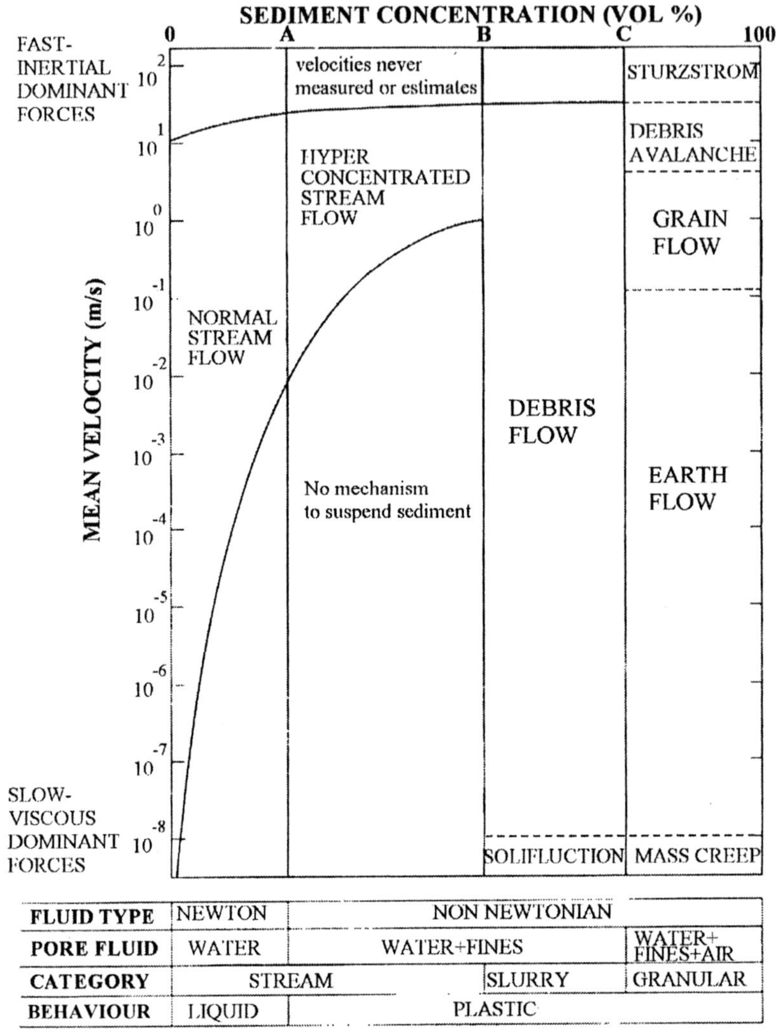

Figure 2.6: Suitable positioning of the terminology that is widely used in practice for hyperconcentrated streamflows in the rheological classification proposed by Pierson and Costa [10].

2.3 The rheological aspects of debris flows

The study of the behaviour of a fluid requires that a suitable constitutive law be defined before any forecasts can be made on a specific flow field. Given the non-Newtonian nature of *debris flows*, summarising this relationship is an extremely complex matter, as it has a wide variety of constituents and, therefore, properties. This means that any attempt at achieving this aim must try to appropriately suit this polymorphic nature. The rheological approach to debris-flow studies aims to provide a macroscopic description of the mixture by means of experiments with artificial channels or those using viscosimeters performed on material samples, that simultaneously measure shear stress and shear rate. The description of the physical behaviour of debris-flows is still a topic of active research; one of the fundamental problems is the characterisation of the material that composes such flows; in other words, the rheology of the materials that constitute a debris-flow. This subject can be handled in two main ways:

1 Considering the entire mass (fluid and solid) as a single material with particular properties. It has been found [9] that a fluid model with yield stress works well for flows with considerable quantities of fine material.
2 Or alternatively, considering the water (sometimes together with the fine fraction) and the coarser material separately. In this case, the models, which are often called grain-flows, are complex and require an in-depth knowledge of the particles and their grain-size.

One difficulty of investigations into the rheology of debris flows is caused by the fact that this kind of flow can contain considerably sized solid particles, which require the use of bulky and costly laboratory apparatus; as the coarse material may be significant in influencing the movement of the mixture, experiments performed on the fine fraction alone, are easier to carry out but unfortunately insufficient. A number of rheological models have been proposed, including Herschel-Bulkley's viscoplastic model and Bingham's plastic model, which are widely used due to their simplicity. It is difficult to say which constitutive equation best represents the behaviour of a granular mass; given the many phenomena that can take place within a lubricated solid mass and the subsequent difficulty of categorising them, they often call for effectively different modelling. It should also be pointed out that the occurrence of the same phenomenon in different geographical areas and with different characteristics has contributed to the development of studies on the question in specific directions, therefore, in China [20] the viscoplastic model is widely used to describe events characterised by large quantities of fine fraction, whereas in Japan [14, 21] the dilatant model is often used to represent those that are characterized by a greater quantity of coarse material. Despite these limits, the continuous improvements in the rheological study of debris-flows make it possible to supply increasingly accurate data on the properties of the mixture, which in turn contributes to improvements in prevention systems and those used to forecast areas at risk from debris-flows.

2.3.1 Rheology of fluids

Rheology is the study of the properties of flow and deformation that bodies are subject to when stress is applied to them. It is therefore the relationship that exists between internal stress and strain. The various materials involved act in different ways; in an ideal gas, stress is always isotropic, in a solid body, deformation is always approximately null, in a linear viscous body, internal stress is directly proportional to the deformation rate (the proportionality coefficient is viscosity) and lastly, in a linear elastic body, internal stress is directly proportional to strain (the proportionality constant is composite stiffness). When a force is applied tangentially to the surface of a body, the action is known as shear stress and the effect that it has on the body is called shear strain. In reference to Figure 2.7, this strain is calculated as $\tan\vartheta = \Delta X / L$.

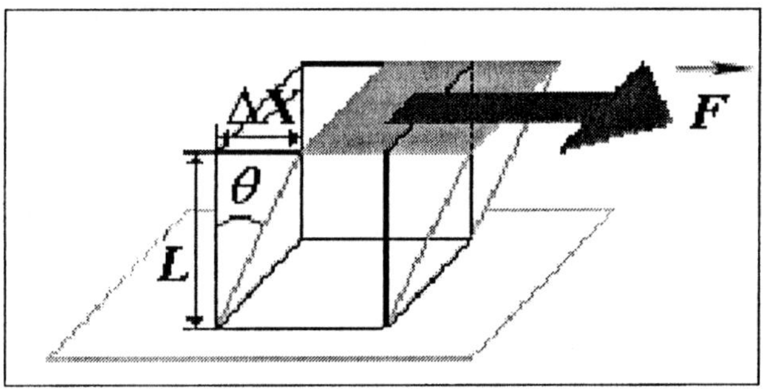

Figure 2.7: Shear stress and shear strain.

When subject to shear stress, fluids tend to deform under shear. On the basis of this property, it could be said that a fluid is a continuous medium in which, in equilibrium, strains are always normal to the respective surfaces, or rather, a medium that cannot support shear stress without deforming due to shear. In a viscous fluid that flows horizontally to parallel layers steadily between two parallel planes, of which the lower one is fixed, two layers with infinitesimal distance dZ between them are considered, A and B are two points that belong to the two layers, respectively, and are positioned vertically one above the other (Figure 2.8). In the layer containing B, the stream flows at velocity v, whilst in that containing A, it flows at velocity (v + dv). The part of the fluid that moves most rapidly exerts shear strain on that below, tending to accelerate it, whilst the slower moving part generates one that tends to slow it down.

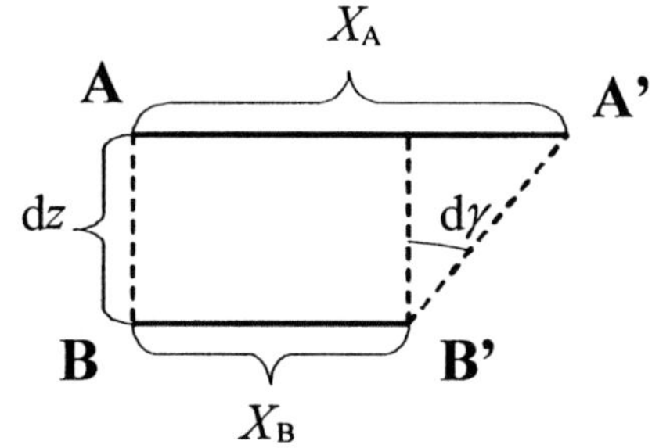

Figure 2.8: Shear strain for an infinitesimal layer of viscous liquid.

This shear strain is proportionate to *S*, the surface area, and *F/S* has the characteristics of a shear strain.

Experience shows that, irrespective of the velocity and thickness of the fluid, shear stress is proportional to the variation in velocity in the *z* direction. Following an infinitesimal interval of time d*t*, point A becomes A' and B becomes B', moreover, the upper layer of the fluid considered is subject to an angular deformation d*γ*, so one obtains:

$$X_A - X_B = (v_A - v_B)\,\mathrm{d}t = \mathrm{d}v\mathrm{d}t = \mathrm{d}z\tan(\mathrm{d}\gamma) \approx \mathrm{d}z\mathrm{d}\gamma$$
$$\Rightarrow \left|\frac{\mathrm{d}\gamma}{\mathrm{d}t}\right| = \left|\frac{\mathrm{d}v}{\mathrm{d}z}\right| \propto \left|\frac{F}{S}\right|, \qquad (2.1)$$

thus the force applied on unit surface is proportional to the rate of angular deformation d*γ*/d*t*.

The relationship that exists between shear stress and strain in a moving fluid can be considered a good representation of the rheological properties of the fluid. It is called the *rheological or constitutive equation* and has an essentially experimental nature, its graphical representation is known as a *rheogram*. In the macroscopic hydrological and geological processes that take place on the Earth's surface, one studies the continuous and irreversible deformation of the material in reply to an applied stress, usually induced by gravity; the material is usually a varied mixture of water, air and solids (rocks, debris, soil).

The response of a material to stress is not merely an intrinsic prerogative of the substance, but also depends on other factors such as temperature, pressure, time and the deformation sequence. Moreover, often in nature, and this is the case for solid–liquid granular mixtures, we encounter multiphase systems, which involve a certain combination of solids, liquids, and gases; the rheological behaviour of these materials also depends on the relative proportion of these components, the grain-size distribution of the solid component and its physical and chemical properties. In fact, in general, those of the fluid and gaseous phase are considered constant.

2.3.1.1 Newtonian fluids

A fluid is called *Newtonian* (curve A, Figure 2.9) if the relationship between applied stress and shear rate is constant:

$$\tau_{zx} = \mu \frac{\partial v_x}{\partial v_z}, \tag{2.2}$$

in which τ_{zx} is the stress exerted in the x direction on the surface normal to z, $\partial v_x / \partial z$ is the shear rate in the direction normal to that of movement and the proportionality coefficient indicated by μ is the so-called dynamic viscosity coefficient (Pa s). For a Newtonian substance such as water, viscosity is constant at constant temperatures and is independent of the shear rate and the time of application of shear stress. Other substances that can be represented as Newtonian are the pure liquids and gases with low molecular weights and homogeneous mixtures of liquids with low molecular weights. All fluids that do not satisfy the law of Newton are obviously defined as non-Newtonian and in such fluids viscosity depends mainly on either the shear rate or time. According to the experiments performed by Fei [22], the critical concentration at which a water–sediment mixture behaves as a non-Newtonian fluid is closely linked to the fine fraction content (diameter 0.01 mm).

When viscosity depends on the shear rate, the previous relationship is no longer valid, but we can write:

$$\tau_{zx} = f\left(\frac{\partial v_x}{\partial z}\right), \tag{2.3}$$

where f is a generic function of the shear rate in the given point; however, the following form is generally used:

$$\tau_{zx} = \mu_a \frac{\partial v_x}{\partial z} \quad \text{with} \quad \mu_a = \frac{f\left(\partial v_x / \partial z\right)}{\partial v_x / \partial z}, \tag{2.4}$$

in which μ_a is the so-called apparent viscosity. It is not usually possible to state the linearity of the previous expression, nor that it represents a physical law, in reality, all the models that seek an explanation constitute semiempirical relationships.

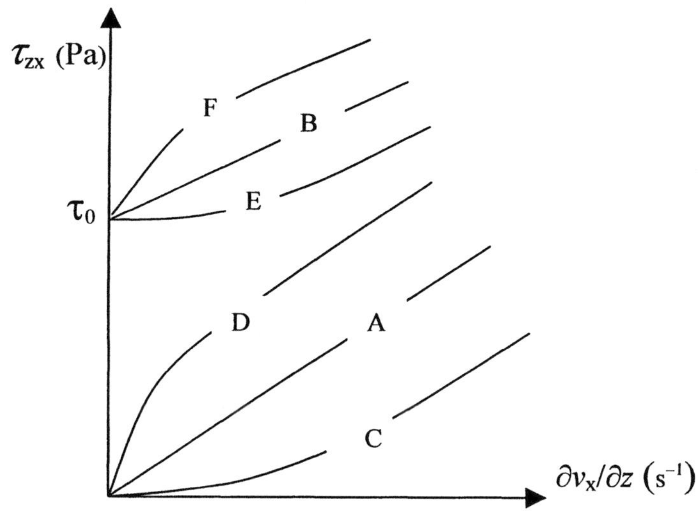

Figure 2.9: Rheological curves for certain fluids (A Newtonian, B Bingham-type fluid, C and D dilatant without and with threshold, E and F pseudoplastic without and with threshold).

2.3.1.2 Bingham's plastic fluids

These fluids require that an initial stress τ_0, known as the limit or yield stress, is exceeded in order for movement to take place, after which, there is a linear relationship between applied stress and shear rate:

$$\frac{\partial v_x}{\partial z} = 0 \quad \text{for } |\tau_{zx}| < \tau_0 \; ;$$

$$\tau_{zx} = \mu_B \frac{dv_x}{dz} + \tau_0 \quad \Rightarrow \quad \mu_a = \frac{\mu_B \dfrac{dv_x}{dz} + \tau_0}{dv_x/dz} \quad \text{for } |\tau_{zx}| > \tau_0 \; , \tag{2.5}$$

in which the coefficient μ_B is known as the plastic viscosity and apparent viscosity diminishes with an increase in the shear rate; graphically eqn (2.5) can be represented by a line with an intercept at τ_0 (curve B, Figure 2.9). Yield stress is a prerogative of fluids in which bubbles or solid particles are suspended in a continuous liquid phase: hyperconcentrated clay and quartz particle suspensions with low shear rates satisfy this empirical relationship.

2.3.1.3 Dilatant and pseudoplastic fluids

In a dilatant fluid (curve C, Figure 2.9) apparent viscosity increases with the shear rate (highly concentrated dispersed systems: starch solutions, potassium silicate solutions and gum Arabic solutions), whereas in a pseudoplastic fluid (curve D, Figure 2.9) apparent viscosity drops as the shear rate increases (concentrated polymer solutions, concentrated emulsions).

$$\tau_{zx} = K\left(\frac{\partial v_x}{\partial z}\right)^n \quad \Rightarrow \quad \mu_a = K\left(\frac{\partial v_x}{\partial z}\right)^{n-1} \quad \begin{array}{l} n > 1 \Rightarrow \text{dilatant} \\ n < 1 \Rightarrow \text{pseudoplastic}, \end{array} \tag{2.6}$$

K is the consistency index and n is the plasticity index. This relationship is also known as the power law and for n=1 and μ=μ_a=K provides the case of a Newtonian liquid. However, generally speaking, the greater the number of rheological parameters introduced in an empirical relationship, the better the equation adapts to experimental data, although the equation itself is more difficult to handle. Due to the many intergranular interaction mechanisms that act in a mixture, the previous equations are not suitable for describing the behaviour of debris flows, for this reason three or more parameter equations are introduced; one of these is the following [4]:

$$\tau_{zx} = \tau_0 + K + \left(\frac{\partial v_x}{\partial z}\right)^n , \tag{2.7}$$

which represents the case of a dilatant fluid (curve E, Figure 2.9) or a pseudoplastic fluid (curve F, Figure 2.9) as the case may be, with yield stress τ_0.

2.3.1.4 Rheopectic and thixotropic fluids

When, in other cases, viscosity depends on time (non steady conditions), there is no longer a univocal bond between stress and strain, however, it varies and depends on the fluid's deformation "history", of which, in a certain sense, it has a memory. There are a number of analysis techniques, however, the most frequently used is that of the closed circuit, which consists in subjecting the fluid to a sequence constituted of an increase and then a decrease in the deformation rate, in order to see how it behaves.

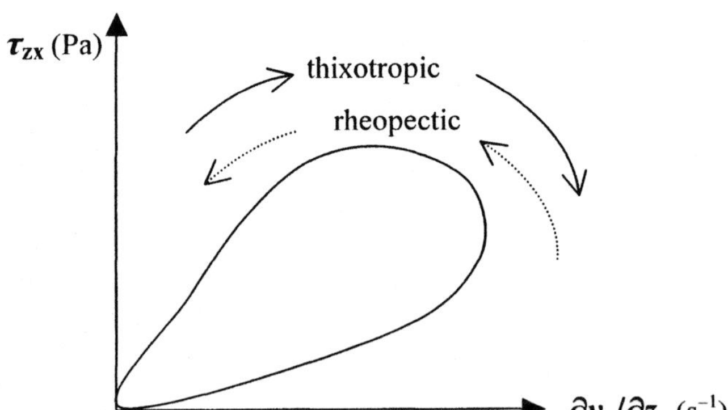

Figure 2.10: Non-Newtonian fluids with time-dependent viscosity.

Using the closed-circuit technique, some fluids called rheopectic fluids show an increase in apparent viscosity with time; for example, bentonite suspensions and certain colloidal solutions behave in this way. Some other fluids called thixotropic fluids, when subject to the deformation sequence for the same application duration of the shear stress, show an apparent viscosity that diminishes with time; this type of behaviour is typical of paints.

2.3.2 Rheology of debris flows

In order to forecast the characteristics of a debris flow it is therefore necessary to start with the determination of its rheological properties. As mentioned previously the direct determination of rheological curves using conventional rheometers does have a problem regarding the presence of considerably sized solid components and, despite the progress that has been made, current apparatus is still inadequate for completely testing debris-flow samples. An indirect method must therefore be used, combining experiments conducted on the finer fraction and theory. One possibility is that of artificially generating different, easily testable suspensions, adding water and coarse particles taken from natural debris-flow samples. (Coussot et al. [23] use this technique on debris-flow samples from the 1995 event that took place at Rio Moscardo in Friuli, Italy). Laboratory observations show that at low concentrations there is a liquid–clay type mixture that shows Newtonian behaviour; as the clay concentration increases, a mixture with a non-null yield stress and deformation is obtained (dilatant or pseudoplastic behaviour with yield stress), thus highlighting considerable temporal properties (thixotropic or rheopectic behaviour).

Sediment concentration and composition play a key role in the choice of a rheological model and for estimating the parameters it includes, such as yield stress or viscosity [24]. In general, Bingham-type models refer to slurry flows, solid–liquid mixtures in which the solid fraction is represented by silt, clays and/or sands, with low deformation values, however, they appear inadequate for describing the behaviour of flows with greater coarse material contents [1]. The influence of larger particles is essential for the rheological characteristics of debris flows and can lead to two types of alteration: on the one hand, it has been observed [25] that by adding larger particles a lower fine fraction content is obtained, and therefore collisions prevail over viscous dissipation and the mixture no longer acts as viscoplastic. On the other, by adding larger particles, the fine fraction content remains sufficiently high to still justify the dominance of viscous dissipation and viscoplastic behaviour. One practical criterion for distinguishing between these is as follows [25]: when the fine and clay fraction (d<40 μm) constitutes 10% of the grain-size distribution, the second type of behaviour occurs and the material is classified as muddy debris; in this case, the most suitable model may be the Herschel-Bulkley model, with an exponent of 1/3, rather than Bingham's model, because the first one works well over a wider range of shear rates ($\partial v_x / \partial z$).

$$\left(\frac{\partial v_x}{\partial z}\right)^n = 0 \qquad \text{for } |\tau_{zx}| < \tau_0$$
$$\tau_{zx} = K\left(\frac{\partial v_x}{\partial z}\right)^n + \tau_0 \qquad \text{for } |\tau_{zx}| > \tau_0 \ , \tag{2.8}$$

in which τ_{zx} is shear stress in the x direction acting on a perpendicular plane to z, τ_0 is the yield stress, K the consistency index and n the flow behaviour index. Certain experiments [23] provide the following values for model parameters with 25% uncertainty: τ_0=2.935 Pa, K=2.190 Pa/s$^{1/3}$ and n=1/3. From these experiments we can also observe that yield stress and consistency index increase considerably (a more or less exponential trend) with an increase in the concentration of coarse particles, whilst the power-law exponent seems to remain constant. However, for granular debris flows, in the absence of a cohesive interstitial fluid, the role of yield stress must be re-examined and models that adopt a friction law and are based on Bagnold-type behaviour, give good results [23]. For example, by starting with a clay mixture and gradually adding coarse particles, for low concentrations, no changes occur because the contacts are lubricated by the high viscosity of the interstitial fluid, and behaviour is, as mentioned previously, viscoplastic; however, when the concentration of the clay fraction falls below 0.10–0.15, the rheological curve drops to a minimum point.

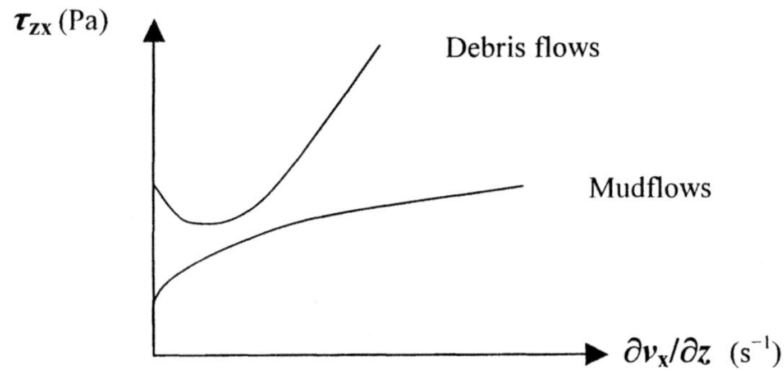

Figure 2.11: Qualitative rheograms for clay and coarse particle mixtures.

In such conditions, actions between the particles, whose submerged weight is sustained by the collisions and by the semipermanent or permanent contacts, cause the movement of the solid fraction. The model that best adapts to these mixtures is a variant of the dilatant model:

$$\tau_{zx} = K\left(\frac{\partial v_x}{\partial z}\right)^2 + \tau_0 \qquad \text{for } |\tau_{zx}| > \tau_0 \ . \tag{2.9}$$

2.3.2.1 Yield stress and effective viscosity

Yield stress is a factor that not only influences debris-flow mobilisation, but is also indirectly connected to the resistance that causes stoppage in the final stages of movement. Yield stress is a characteristic property of solid bodies and represents a fundamental concept of rheology. It can be used to describe debris-flow behaviour before movement and during the initial stages of deformation, under the action of external forces, when the transition from a solid body to fluid mass takes place. However, during experiments one can observe that yield stress is always higher than the corresponding values in stoppage conditions [26]. Yield stress therefore presents a certain form of hysteresis, which must be considered in forecasting the overall distance covered by a debris-flow (*runout*) for a given topography. In fact, by using the yield stress relative to movement triggering, in a runout-distance forecasting model, one is likely to seriously underestimate it. Moreover, in studying hysteresis, it must be considered that, being a process guided by gravity (and therefore by active shear stress), the phenomenon in the field cannot always be well represented by rheological curves derived from laboratory measurements using rotational apparatus, in which the process is instead guided by the shear rate [27].

For the direct calculation of the yield stress τ_0 in debris flows, Johnson's formula is often used [9]:

$$\tau_0 = \rho_f g h_0 \sin\theta \,, \tag{2.10}$$

where ρ_f is fluid density, h_0 the asymptotical depth of the current and θ the mean inclination of the slope.

In a solid–liquid mixture with particles smaller than 0.01 mm, and therefore highly cohesive, at relatively low concentrations, the distance between the particles dispersed in the fluid is so large that the influence of one particle on surrounding ones is negligible and in each point of the fluid the effect of all the particles can be evaluated by adding the effects generated by each one together. However, when the concentration exceeds a threshold value, the mixture no longer behaves as Newtonian, but the appearance of yield stress justifies the adoption of a Bingham-type model. This threshold concentration varies with sediment type and its mineral composition. In his investigations into the minimum concentration that would lead to the appearance of a yield stress, Fei [22] indicates τ_{0lim} =0.5 N/m^2 as the critical condition for distinguishing a Newtonian fluid from a Bingham-type fluid (if the yield stress exceeds the proposed limit value). Moreover, the minimum concentration is closely linked to the fine particle content (dimension smaller than 0.01 mm). For example, for a sediment mixture of which 70% is constituted by fine particles, the minimum concentration of solid material inside the mixture, for a yield stress to appear, is 4%. Starting from such assumptions, Fei's study [22] developed expressions that can be used for the evaluation of relative viscosity μ_r and yield stress τ_0 (N/m^2) in water–sediment mixtures, in which the fine particle content is negligible (smaller than 0.01 mm) and assuming the relative particles are uniform and spherical:

$$\mu_r = \frac{\mu}{\mu_0} = (1 - 1.35C_v)^{-2.5} \tag{2.11}$$

$$\tau_0 = 9.9 \times 10^2 \exp\left(8.45\frac{C_v - C_{v0}}{C^*} + 1.5\right), \tag{2.12}$$

in which μ is the viscosity of the interstitial fluid, μ_0 the viscosity of pure water at the same temperature as the fluid, C_v the volumetric concentration of the sediment, C^* is the maximum volumetric concentration of the sediments and C_{v0} is the critical concentration of the sediments for the appearance of non-Newtonian behaviour in the mixture, which can be calculated with $C_{v0} = 1.26(C^*)^{3.2}$. Moreover, such suspensions would behave as Newtonian fluids even at low solid concentrations (C_v <0.27), whilst a Bingham-type yield stress would appear for solid concentrations higher than 3.5; this yield stress, which is much lower than the case of fine suspensions, would reflect the friction resistance due to mutual collisions between particles [4].

A parallel study conducted by O'Brien and Julien [28] at low shear rates in laminar-flow conditions without secondary flows, shows that yield stress τ_0 and viscosity μ, in the ambit of Bingham's model, increase exponentially with the volumetric concentration C_v of the solid fraction including clay, silt and sand sediments, but not coarser clastic material:

$$\tau_0 = \alpha_1 e^{C_v \beta_1} \tag{2.13}$$

$$\mu = \alpha_2 e^{C_v \beta_2}. \tag{2.14}$$

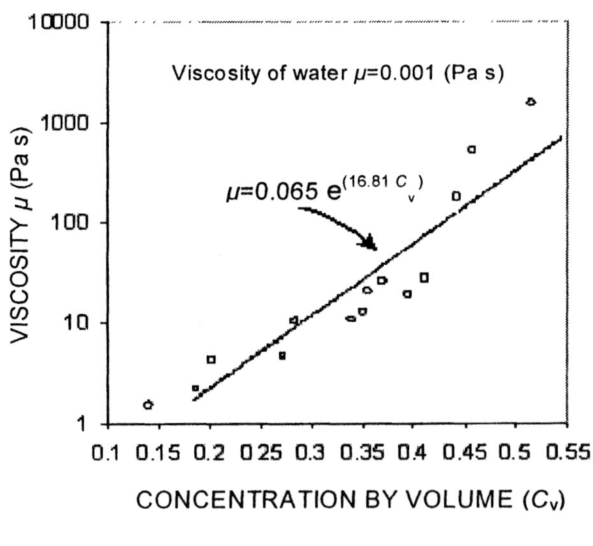

Figure 2.12: The exponential empirical relationship between viscosity and solid concentration obtained in the rheological tests of O'Brien and Julien [15].

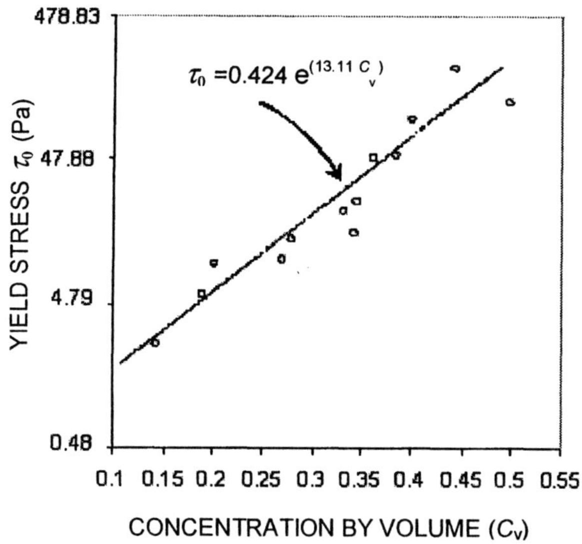

Figure 2.13: The exponential empirical relationship between yield stress and solid concentration obtained in the rheological tests of O'Brien and Julien [15].

In eqns (2.13) and (2.14) the constants α_1, α_2, β_1 and β_2 are calculated in laboratory experiments, and when α_1=0.424 Pa, β_1=13.11, α_2=0.065 Pa s and β_2= 16.81 the previous correlations are shown in Figure 2.12 and Figure 2.13.

One can observe the wide variability of the two parameters with sediment concentration: these may vary by three orders if the concentration of the sediments passes from 10% to 40% [28]. At low solid concentrations, the effect of adding coarser material on the yield stress is practically negligible and yield stress follows an exponential law such as eqn (2.13); one could therefore say that for high total solid concentrations, yield stress might not be particularly high; this would be physically acceptable by referring to the intergranular contacts lubricated by the water–clay interstitial fluid [28]. The movement of the coarser particles could therefore continue even at low deformation rates, considering that the intergranular contacts responsible for energy dissipation, would be continuously lubricated by interstitial fluid. However, the mixture behaves differently when granular material increases, as yield stress grows much more rapidly, the mixture fractures easily and does not consent to continuous deformation; the increase in direct contacts between grains causes greater energy dissipation, in contrast to the previous case with lubricated contacts, at least at low deformation rates. Moreover, if there is a high clay content, the above-mentioned rheological properties cannot be simply characterised by the volumetric concentration of the solid particles alone, but the pH, salinity, hysteresis, and temperature must also be introduced, thus leading to the adoption of specific indexes such as that of liquidity suggested by Locat [29]. This underlines the difficulties in objectiveness, even from an experimental point of view, in describing the rheological behaviour of mixtures composed of water, clay and solid material and in identifying for them constitutive laws that can be subsequently used for modelling solid–liquid flows, in order to create the basic fundamentals for a qualitative forecasting of their behaviour on the field.

Chapter 3

Debris-flow phenomenology

3.1 Introduction

Debris flows occur when poorly sorted sediments, shaken and mixed with water, flow down a slope under gravitational force, deforming irreversibly and maintaining a free surface. One condition for the occurrence of this phenomenon is therefore the presence of unconsolidated material and rocky fragments, combined with a vast, continuous or intermittent supply of water and vegetation distributed along the slope. Debris-flows are considerably and synergically influenced by the solid and fluid forces that helps distinguish them from other phenomena in which one type of force is clearly dominant over the other [3]. As well as providing the basis for a physical classification of such phenomena, this characteristic also accounts for their immense destructive power. Even on gentle slopes, debris-flows have a sufficiently fluid consistency to travel considerable distances, and their impact can therefore be extremely widespread. Most debris-flows originate from a static and practically rigid mass of sediments that are mixed with water. When it starts to move, the mixture has the same characteristics as a liquid flow and on deposition it reverts to its initial characteristics. Debris-flows are usually caused by the collapse of portions of hill or mountainsides and can be the natural evolution of a landslide, slump or debris avalanche, after heavy rainfall or earth tremors [21]. The erosive nature of a debris-flow itself represents a factor that favours the phenomenon. Whereas certain types of mass wasting are able to erode up to a few metres below the channel, debris-flows can exceed 10^9 m^3 in volume and release 10^{16} J of potential energy, however, even a 10^3 m^3 event can cause serious damage. Flow is unsteady and nonuniform and can support itself for more than 10^4 s; peak velocities can reach as much as 10 m/s, which makes inertial forces particularly important in its dynamics.

Figure 3.1: The propagation of a wave during the debris flow at Tahoma Creek from Mount Rainier, July 26, 1988 (photo credit: U.S. Geological Survey/photo by G.G. Parker).

Sediment concentration by volume tends to be greater than 50%; in the head in particular it can vary from 30–65% and generally drops towards the tail. Streamflow can vary in depth from 1 m to several meters and mean channel inclination ranges from 40° in origin areas to 3° in deposition areas [3].

This chapter presents a study backed up by thorough bibliographical research performed on field observations and recordings, laboratory experiments and deposit analysis, which attempts to create a phenomenological framework of this complex phenomenon.

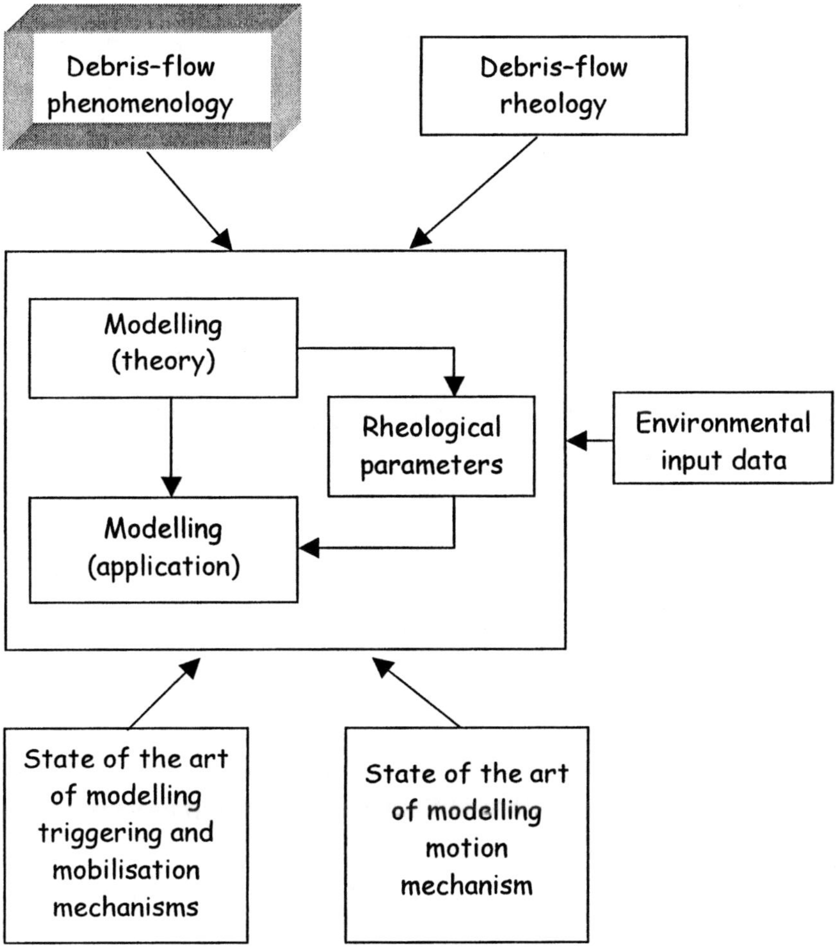

Figure 3.2: In the chart, showing the debris-flow study approach adopted in this book, the grey frame highlights the field of analysis concerning debris-flow phenomenology, the subject of this chapter.

3.2 Typical debris-flow environments

Debris flows mainly characterise three types of environments:

- Semiarid areas;
- Alpine areas;
- Volcanic areas.

In semiarid areas, debris flows are caused by brief but intense pluviometric events where heavy rain falls in a relatively small basin, causing an increase in the intergranular pressure of the ground and therefore the mobilisation of superficial debris over relatively stable bedrock.

In the second case, the phenomenon is usually caused by snow melting, but scarce vegetation and a noncontinuous water supply are equally favourable to occurrence.

In volcanic regions the phenomenon is better known as a *lahar*, and takes place during or after a volcanic eruption. According to Costa [30], strong rainfall, glacier melting, pyroclastic flows with water and earthquakes are potential lahar-triggering mechanisms.

In any case, debris flows are the result of some kind of hill– or mountainside collapse; as it descends along a slope, the landslide becomes moist due to fluidification or dilatation, thus increasing its mobility.

3.3 Debris-flow characteristics

Field studies and recordings, laboratory experiments and deposit analysis permit us to draw up an albeit empirical phenomenological framework of this complex phenomenon.

3.3.1 Composition and grain-size

The appearance of debris flows is commonly compared to that of wet cement, however, its composition varies considerably within the context of what is known as a dispersed, scarcely sorted, multi phase flow that is mixed with air and water [3]. The rheological properties of debris-flows are closely linked to their solid and water content and their particularity lies in the fact that they are able to flow even with small quantities of liquid. Particle size can vary from fine material to large clasts and the particles are dispersed within a flow that can include any material encountered along its path (Figure 3.3).

In Japan, a boulder weighing some 3000 tonnes was observed to have been carried several kilometres by a debris flow [14]. A slurry occurs when high concentrations of silt and clay are mixed with water to form a viscoplastic flow; if particles of sand are present, the flow is known as muddy, or alternatively, a mixture may be composed of rocky fragments dispersed within a flow composed of fine particles and water. The density may vary from 1400 kg/m^3 to 2530 kg/m^3. Grain-size distribution, grading curves, the proportion of the fine fraction to the coarse one, water content and shear rate characterise the composition and the rheological properties of a debris-flow (viscosity, yield stress, density and concentration).

Debris flows frequently interact with their surrounding environment through continuous erosion and deposition processes and therefore have a stochastic and chaotic nature, which makes evaluating their typical constituents all the more difficult.

Figure 3.3: Photograph showing the presence of large boulders in the flow (debris-flow associated with the December 1999 storm in coastal Venezuela / Photo credit: U.S. Geological Survey).

Following grain-size analysis performed on various samples from debris- flow deposits, Znamensky and Gramani [31] subdivided the solid fraction into three classes (Table 3.1): micrometric (μ), mesometric (mm) and macrometric (m). These classes are in close correlation with viscous, muddy and stony debris-flows, respectively, as defined by Julien and O'Brien [32] or Takahashi [33].

A prevalent fine granule content will usually cause a slurry and the resulting debris flow will have a viscous behaviour, whereas if the fraction composed of larger particles prevails, it is probable that the debris-flow will be muddy and will therefore have dispersive/dilatant behaviour. The reduction in granule size in the solid fraction during runoff is connected both to collisions, in an inversely proportional way to volumetric concentration, and to the friction contacts between the granules and the channel, in a directly proportional way, thus contributing to modifying the rheological properties of debris-flows during the event. This process of grain dimension reduction during a debris-flow can be described for spherical granules by an exponential law [31], such as

$d = d_0 \cdot e^{-(L \cdot C_{GR}/3)}$, which expresses the final diameter d (mm) as a function of the initial diameter d_0 (mm), runout distance L (km), the distance covered by a debris-flow before stoppage and measured in a straight line, and a coefficient of volumetric reduction C_{GR} (km^{-1}), particular to each grain-size class, calculated during experimentation.

Table 3.1: Grain-size classification of the solid fraction of a debris flow proposed by Znamensky and Gramani [31]; d is mean diameter of the granules of the solid fraction in question.

CLASS	***GRAIN–SIZE RANGE***: *magnitude and description*			
	From		**To**	
	d (mm)	*description*	*d (mm)*	*description*
Micro range (μ)	0.00024	Very fine clay	0.062	Very fine sand
Meso range (mm)	0.062	Very fine sand	16	Coarse gravel
Macro range (m)	16	Coarse gravel	4000	Rock blocks

3.3.2 Physical behaviour

The development of debris-flows involves three areas: the area of origin of the sediment pile in the uphill section of the basin, which is usually a landslide area; the sediment transport zone, involving channels that develop along very steep slopes and where erosion and deposition processes are more or less in equilibrium, and lastly, the deposition area [15]. Debris flows have a distinctive way of moving as a succession of waves with steep and lobed heads, rich in considerably sized clasts. Initially, these waves occur at intervals that vary from a few seconds to every ten minutes or so, however, this frequency subsequently drops to every few hours. This mechanism would seem to depend on the periodical changeover of debris in the area of origin, temporary channel blockages, bank erosion, and the friction between the flow and the channel bed.

3.3.3 Velocity and slope angles

Typical debris-flow velocities range from 0.5 m/s to 20 m/s: a somewhat vast interval that can be attributed to variations in flow composition (sediment concentration and size) and slope geometry (inclination, dimension and meanderings). Velocity may also vary within the same event and, in particular, from the area of origin to that in which the material is deposited, where deceleration and progressive deposition considerably reduce the intensity of the flow. Figure 3.4 shows the velocity observed in a number of debris-flow events.

LOCATION	VELOCITY (m/s)	SLOPE (%)	DENSITY (g/cm^3)	VISCOSITY (poise)	%CLAY	DEPTH (m)	SOLIDS (% Wt)	REYNOLDS NUMBER	REFERENCE
Rio Revntado, Costa Rica	2.9–10	4.6–17.4	1.13–1.98	–	1–10	8–12	20–79	–	Waldron, 1967
Hunshui Gully, China	10–13	–	2–2.3	15–20	3.6 (<0.05 mm)	3–5	80–85	40 000	Li & Luo, 1981
Bullok Creek, New Zeland	2.5–5.0	10.5	1.95–2.13	2100–8100	4	1	77–84	28.57	Pierson, 1981
Pine Creek, Mt. St. Helens	10–31.1	7–32	1.97–2.03	200–3200	–	0.13–1.5	–	200	Fink et al.,1981
Wrightwood, Canyon Ca. (1941)	1.2–4.4	9–31	2.4	2100–60 000	<5	1.2	79–85	23.8	Sharp & Nobles, 1953
Wrightwood, Canyon Ca. (1969)	0.6–3.8	9–31	1.62–2.13	100–60 000	–	1	59–86	1.33	Mortan & Campbell, 1974
Mayflower Gulch, Colorado	2.5	2.53	2.53	30 000	1.1 (<0.04 mm)	1.5	91	3.2	Curry, 1966
Dragon Creek, Arizona	7.0	5.9	2.0	27 800	–	5.8	80	29.2	Cooley et al., 1977
Jian–Jia Ravine, China	8.0	0.06	2.3	15.5–1736	–	1.4	89	148–11 561	Li et al., 1983

Figure 3.4: Observations made of the physical properties of a number of debris flows [1].

The possibility of high velocities could suggest the influence of turbulence on the flow, however, authors in the literature express conflicting opinions on the matter. Bagnold [7] observes that the presence of clay and silt particles should reduce turbulence, a view that is shared by Johnson [9], who adds that turbulence itself is solely associated with the narrowing of the debris-flow pathway or sudden changes in level.

However, following direct observations, Pierson [34] notes that with a change in velocity from 3 to 5 m/s, the flow becomes turbulent despite being composed, to a large extent, by clay and silt particles, the same behaviour is also observed when the concentration in weight of the sediments nears 70%; however, no specific criteria are proposed that allow us to define these transition boundaries with certainty.

Debris flows usually take place along particularly steep slopes, nevertheless, the particular fluidity of the phenomenon makes flow possible even along much gentler slopes; typical inclinations observed range from 15% to 32%, whereas the area in which stoppage occurs is characterised by gentler inclinations that do not exceed 3% [14].

3.3.4 Mixture properties

Whereas some debris-flow characteristics depend closely on the dynamic character of the phenomenon, others can be easily evaluated in static conditions, including the grain-size distribution of the mixture's solid fraction. Data on this aspect shows that within a single debris-flow it is possible to encounter grains that range in size from clay to large boulders, which, as is well known, are difficult to sample and are consequentially often neglected in measurements. The majority of the mass is obviously composed of sand, gravel and slightly larger granules, with silt and clay size particles making up less than 10%. On the one hand, this first piece of information on the wide variety of mixture components immediately highlights the limitations of theoretical models, which generally assume a single grain-size or the preponderance of the fine fraction and on the other, it supports the idea that grain-size variation is a critical characteristic for debris-flow behaviour.

Rheometric experiments performed by gradually adding water to samples taken from debris-flow deposits, have shown that the behaviour of the mixture varies considerably even with small variations in the solid fraction (concentration), deformation rate (a parameter used for estimating kinetic energy) and grain-size distribution (in particular, the silt and clay content, which, as illustrated by the experiments of O'Brien and Julien [28], has a strong influence over solid–liquid interactions). The analysis of such trends has led to the development of a series of studies based on the analogy between granular mixtures and dense and ideal gases [35]. In dense gases, the concentration of the different molecular species, their kinetic energy (temperature) and their forces of interaction combined determine the overall properties of the mixture, such as density, resistance and changes in state. In exactly the same way, debris-flow properties depend essentially on concentration, kinetic energy and the interactions between the solid and liquid components; this has led to an increasing preference for study approaches that focus on the distinctive particularities and the interactions of the two components, thus foregoing others based on the simplified hypothesis that within a debris-flow, the solid and liquid phases are inseparably linked to form a monophase substance.

The salient mechanical properties of a solid particle are:

- density ρ_s
- typical diameter d, defined as the diameter of a sphere of equal volume.
- the friction coefficient $\tan\phi$, in which ϕ is the angle of sliding friction, a function of the shape and roughness of the grain.
- the restitution coefficient e that varies between 1, corresponding to a perfectly elastic granule, and 0, corresponding to a perfectly inelastic granule.

The solid component occupies a fraction v_s of total mixture volume and has a distribution of d that, as mentioned above, can span various orders of size.

Similarly, the fluid component is characterised by the corresponding properties:

- density ρ_f (usually lower than that of the solid component)
- effective viscosity μ
- volumetric fraction v_f .

For the normal mean stresses typical of debris flows (<100 kPa) the solid and fluid components cannot be compressed and the variations in v_s/v_f are far greater than those of ρ_s/ρ_f.

Two additional properties connect the behaviour of the solid and the fluid: the fractions in volume satisfy the condition for which $v_s + v_f = 1$ (therefore the density of the mixture is expressed by $\rho = \rho_f v_f + \rho_s v_s$), moreover, the intrinsic hydraulic permeability k represents the resistance to relative motion of the two phases. Intrinsic hydraulic permeability represents the properties of the soil to be crossed by the water with a vertical downward movement and remains indirectly defined by the law of Darcy, a phenomelogical-type relationship that describes the monodimensional filtration motion in natural, isotropic and homogeneous soils, in which the regime can be considered laminar, by means of the linear relationship between specific discharge and hydraulic gradient:

$$v_f = -\frac{k}{\mu} \cdot \nabla(\rho g z + p), \tag{3.1}$$

which includes filtration velocity v_f, otherwise known as specific discharge due to its being equal in modulus to the relationship between the discharge and the overall area, vacuums plus full parts, of the section normal to the streamflow, the hydraulic gradient $\nabla(\rho g z + p)$, in which p is pressure, ρ density, g gravitational acceleration and z the altitude of a generic point in relation to an arbitrary reference plane, which is positive if directed upward, the viscosity of the medium μ and lastly, the intrinsic hydraulic permeability k [L^2] that depends on the microstructural composition of the medium alone. In natural soils, the evaluation of k is somewhat uncertain, due to the complex geometry of the medium and the distribution, dimension, and shape of the particles that make it up.

Table 3.2: Typical values for the basic physical properties that characterise a debris flow [3].

SOLID PROPERTIES	SYMBOL	TYPICAL VALUES
Mass density (kg/ m^3)	ρ_s	2500–3000
Mean diameter (m)	d	10^{-5}–10
Friction angle (°)	ϕ	25–45
Restitution coefficient	e	0.1–0.5

FLUID PROPERTIES	SYMBOL	TYPICAL VALUES
Mass density (kg/ m^3)	ρ_f	1000–1200
Viscosity (Pa s)	μ	0.001–0.1

MIXTURE PROPERTIES	SYMBOL	TYPICAL VALUES
Solid volume fraction	v_s	0.4–0.8
Fluid volume fraction	v_f	0.2–0.6
Friction angle (°)	ϕ	25–45
Hydraulic permeability (m^2)	k	10^{-13} – 10^{-9}

By centring analysis on the distinguishing properties of the solid and fluid components, with a view to the application of formal theories on solid–liquid mixtures, we must define exactly what constitutes the interstitial fluid, given that a debris flow can contain very fine particles in solution or suspension. One simple criterion that can be useful in distinguishing between the solid and fluid fraction can be proposed in reference to a suitably chosen characteristic dimension; for example, using d' to indicate the characteristic dimension taken as a reference, it can be assumed that all grains with a diameter $d < d'$ act as an integral part of the fluid and all those with diameter $d > d'$ are an integral part of the solid. Alternative criteria based on particle sedimentation times have been proposed in order to overcome the arbitrary nature inherent in the choice of the characteristic diameter d' [3]. If the duration t of a debris-flow event is sufficiently long in comparison to the sedimentation time of a granule with a diameter d immersed in pure water of viscosity μ_w in static conditions, the granule should be considered part of the solid fraction; its suspension within the debris-flow is a consequence of the interaction with the other particles and the turbulence of the fluid (Figure 3.5). If, as a consequence of the viscous resistance of the water alone, a granule can remain suspended for periods of time longer than t, we can assume that the granule acts as an integral part of the fluid fraction. Characteristic debris-flow durations range from 10 s for smaller scale events to 10^4 s for more extensive ones. A granule's sedimentation time can be estimated by dividing half of the mean flow height ($h/2$) by the granule's sedimentation velocity v_s, which can be calculated using the law of Stokes or a more general equation that considers the inertial effects.

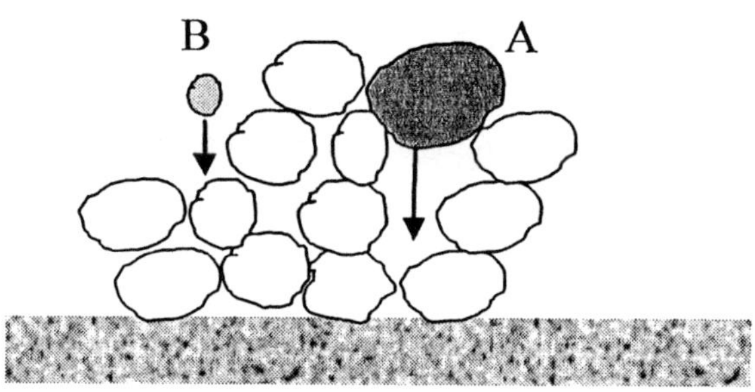

Figure 3.5: Diagram showing the distinction between a small granule B, which, being lighter and whose total weight is diminished by Archimedes' buoyant force, is counterbalanced by viscous forces alone, which cause it to remain suspended, and a larger granule A, whose suspension also requires interaction with the other particles [3].

According to this criterion, if $h/2v_s < t$ the particle pertains to the solid fraction. The semiheight for real debris flows ($h/2$) is usually between 0.01 m and 10 m depending on the entity of the event; thus, in the case of small-scale events ($t \approx 10$ s and $h/2 \approx 0.01$ m) it is $h/2t \approx 0.001\,\mathrm{m/s}$, and therefore, if the sedimentation velocity is $v_s < 0.001$ m/s, the particle acts as part of the fluid. It has been discovered [36] that similar sedimentation velocities in water correspond to granules with diameters of less than 0.05 mm, giving an estimate for the critical diameter, which is practically on the boundary between that of sand and that of silt (approximately 0.0625 mm). In such cases, the Reynolds number ($N_{RE} = d\rho_f v_s / \mu_w$, in which μ_w and ρ_f are the viscosity and density of water) is much lower than 1, thus confirming that viscous forces dominate over inertial ones for this type of granule. In conclusion, a general guideline can be obtained for distinguishing between the solid and liquid fraction, by assuming that granules larger than silt are part of the solid component of a debris-flow, whereas smaller ones act as part of the fluid.

If we assume that the fluid fraction of a debris flow is also composed by a fine fraction, in line with points made previously, the definition of the density of the fluid ρ_f must be altered in order to obtain a property that suitably describes the model adopted:

$$\rho_f = \rho_s \nu_{fine} + \rho_w (1 - \nu_{fine}), \tag{3.2}$$

in which ν_{fine} is used to indicate the volumetric fraction of the fluid occupied by the fine fraction (silt and clay), ρ_w is the density of the water and ρ_s that of the fine grains, which, for simplicity's sake is assumed to be equal to that of the larger granules. Comparisons between direct measurements of ρ_f and values calculated using eqn (3.2) show an error of 10% [3].

The presence of a fine fraction within the interstitial fluid not only influences its density, but also effective viscosity; Einstein [37] was the first to formulate an equation to describe the effects of the presence of fine particles on the viscosity of a fluid:

$$\mu / \mu_w = 1 + 2.5\nu_{fine}, \tag{3.3}$$

in which μ is the effective viscosity of the suspension and μ_w the viscosity of the water. This expression is applied to diluted suspensions of chemically inert spheres with $\nu_{fine}<0.1$ and $N_{RE}<<1$, and although it can be a useful approximation for real debris flows, in certain cases, especially when the fine fraction is greater, an extension is required. The various formulae that have been proposed for forecasting viscosity at higher fine-particle concentrations, include the empirical relationship of Thomas (1963), which is valid for suspensions in which Archimedes' thrust and resistance dominate the forces of solid–liquid interaction:

$$\mu / \mu_w = 1 + 2.5\nu_{fine} + 10.05\nu^2_{fine} + 0.00273\exp(\nu_{fine}), \tag{3.4}$$

which is reduced to Einstein's formula for low concentrations. The downfall of these formulae is that they neglect the effects of the deformation rate, which are considerable if $\nu_{fine} > 0.4$, if the geometry of the grains is very nonspherical, or if the Van der Waals' electrostatic bonds between clay and colloidal particles become significant.

3.3.5 Flow profile

Debris-flow dynamics can be divided into three distinct time phases. The first is a brief initial phase in which the density of the mixture differs very slightly from that of water, the solid particles do not interact and rapid "saturation" takes place. This is followed by a longer intermediate phase, which corresponds to the developed motion of the debris-flow, in which density is particularly high and interactions become predominant. In the third and final phase, the process ends following deceleration, thus resulting in the disintegration and the deposition of the debris and the formation of a rigid skeleton, corresponding to the consolidated solid fraction, through which the water is able to filter out. Although formulating a single model that accurately describes all three phases is an extremely complicated task, it is, however, possible to refer to the median process as a good approximation [38]. A typical debris-flow wave, photographed within this timeframe, can be spatially divided into three main regions, as shown in Figure 3.6. The anterior section of the flow (also known as the snout) forms the head and usually has a maximum relative extension of ten metres; it has a forwards protuberation composed mainly of rock fragments due to the inverse grading phenomenon (see Section 3.4.4). Takahashi [21] justifies this conformation with the observation that, being full of large boulders, the head tends to slow down during descent and increase in volume. The "head effect", or rather the entire set of its distinctive characteristics, can be described by the following dynamics [39]:

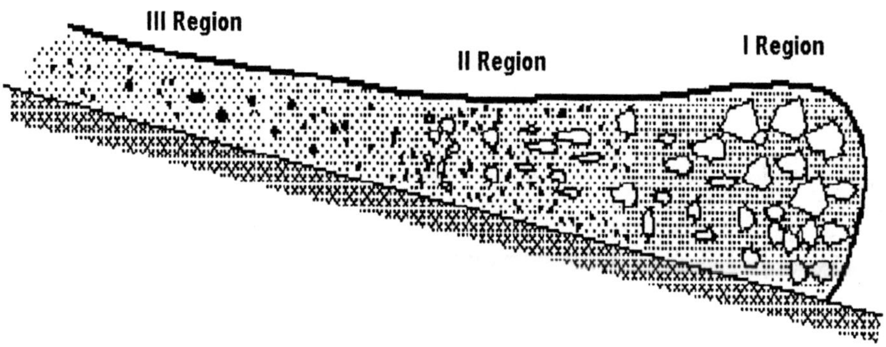

Figure 3.6: The various regions within a debris-flow wave.

- Local increase in the coarse-granular fraction, which, in addition to the inverse-grading mechanism, is also due to the incorporation of any object it meets on its path (bridges, cars, trees, etc.).
- Local reduction in water content: the intergranular pressure calculated at the base of the head is practically null, whereas in the rear section it is usually equal to or greater than that required to balance total normal stress and to suspend the granular mass.
- Reduction of the total flow discharge.
- Local reduction in the velocity that causes a slight slowdown in the oncoming flow and in some cases, leads to stoppage.

There are three main mechanisms responsible for this type of behaviour: the loss of water and the fine fraction, due to adhesion to the channel walls; inverse grading, which, in accumulating large boulders at the head, favours the increase in friction interactions in the contacts between them and the possible formation of dams and blockages and, thirdly, the plastic nature of the interstitial fluid, which also contributes to further slowing of the flow.

The second region (commonly known as the body), characterised by an increasingly tapered flow, is composed of a turbulent mixture of water, fine particles and sediment transported along its course, whereas the last region, which is slender and much more diluted than the previous two, has a high mud concentration and continues to flow long after the wave has passed.

3.4 Salient aspects of debris-flow phenomenology

Some of the characteristics that have been observed in debris flows are central to the mechanics of the phenomenon, and should therefore be suitably considered in any attempts at modelling. These characteristics are: dilatancy, particle segregation (still to a large extent misunderstood on a theoretical level) and the mechanisms that govern grain dispersion throughout the flow.

3.4.1 Dilatancy

Reynolds (1885) was the first to give a definition of this property, particular to granular mixtures. If a system of compact spherical particles is subject to a load and therefore, deformation, purely geometrical considerations suggest that the particles must slide over one another, thus causing an overall increase in volume. This phenomenon also takes place in granular systems due to high deformation rates, and is induced by intergranular collisions and the consequential dispersive pressure [7]. Dilatancy in granular materials occurs accompanied by the effect of normal stress, which tends to increase the thickness of the deformed layer. Debris flows show dilatancy because, under constant environmental pressures, deformation causes an expansion in the interstitial space [40].

3.4.2 Internal friction and cohesion

Debris flows have a behaviour that is midway between that of a solid and that of a liquid. From a solid point view, debris can be stacked and the free surface of the pile has a maximum inclination corresponding to the material's angle of repose. This behaviour, caused by friction and the possible cohesion between grains, is described by Mohr–Coulomb's limit criterion, which establishes that the critical condition is reached on a surface when shear stress assumes the value:

$$|\tau| = c + \sigma \tan(\phi), \tag{3.5}$$

where c is cohesion, σ normal stress and ϕ the angle of internal friction, which depends on the angle of internal friction of the single grains and the geometry of the debris pile. In static conditions, this usually varies from 20° to 45°, whereas it drops to 3−4° in dynamic conditions under high deformation rates, which do not, however, preclude friction contacts to the granules. Generally speaking, the behaviour of a material can be said to be plastic if shear rate and stress are independent, viscous if there is any type of dependence between the two variables and viscoplastic if both characteristics are present; from this point of view, debris-flows can be described as viscoplastic, given the evolution that takes place from the areas of origin to the stoppage fans.

3.4.3 Fluidification

The term fluidification refers to the process by which, given an initially static bed of solid particles, over which an ascensional water flow passes, the solid phase gradually develops a semifluid behaviour, following the reduction of internal friction and cohesion; in such conditions the bed is said to be fluidised. When a water flow flows across a static bed of particles, a pressure gradient develops throughout the bed and the balance of forces acting on a single particle alters due to the appearance of drag force (the force exerted by the fluid on the solid in the direction of the flow, which, in the case of an ascensional flow, is upwards). As drag force depends on streamflow velocity (see Section 4.3.1), an increase in velocity also causes an increase in the entity of the drag force exerted on the particles, thus reducing the resistance of the system and causing the bed to expand. Whereas in a static bed the particles are in direct contact with one another and the friction forces balance the weight force, in an expanded bed, the particles move apart under the action of drag force. Fluidification takes place when the force associated with the pressure gradient through the bed exactly counterbalances the net weight of the system (weight of the solid pile reduced by Archimedes' force):

$$\Delta p A = (1-\varepsilon)\cdot(\rho_p - \rho)\cdot gAH\,, \tag{3.6}$$

in which A is the transversal area of the bed, ε is the fraction of vacuums or porosity (ratio of the volume of the liquid phase to the overall volume of the mixture), ρ_p and ρ are, respectively, the density of the solid particles and that of the liquid, g is gravitational acceleration and H the height of the bed. In steady conditions, the total mass of the solid remains constant and equal to $M = (1-\varepsilon)\rho_p AH$; in addition, by considering ρ_p and A constant, between two generic instants 1 and 2 of the process, porosity and the height of the bed could vary, but in such a way that $(1-\varepsilon_1)H_1 = (1-\varepsilon_2)H_2 = \text{cost}$, thus keeping the total mass of the solid constant.

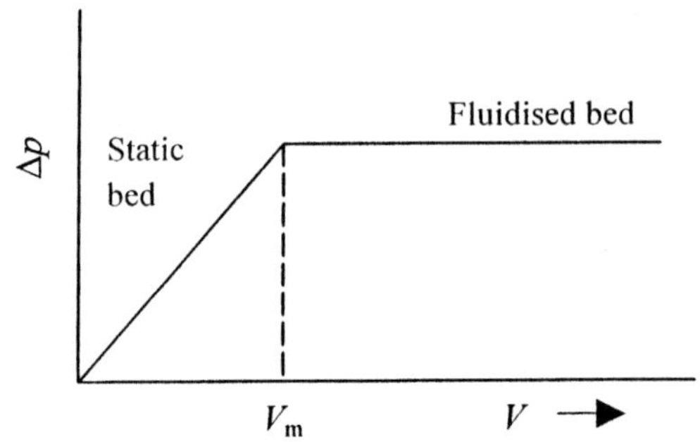

Figure 3.7: The trend of the pressure gradient Δp relative to the velocity of the streamflow V in the transition from static bed to fluidised bed.

Moreover, it can be assumed that for small changes in pressure, fluid density remains more or less constant. In short, the second member of eqn (3.6) is constant, which gives the pressure gradient in a fluidised bed as constant. The fluidification process therefore depends on streamflow velocity. In particular, there is a minimum velocity V_m beneath which the bed continues to behave as a static accumulation of cohesive particles; however, when velocity exceeds V_m, the bed expands and the particles tend to move away from one another as porosity and bed height increase, because the pressure gradient must remain constant during fluidification (Figure 3.7).

As a consequence of fluidification, large masses of rock, soil and snow can cause shallow flow on surfaces much less inclined than the angle of the repose of the material and consequentially flow long distances. As shown previously, the first cause of such behaviour is the presence of a water flow, which, as it runs across the bed of sediments and develops a pressure gradient, causes its subsequent dilation. The second cause is connected to the fact that during movement, granular mixtures behave like non-Newtonian fluids and, when subject to high deformation rates, they develop considerable dispersive pressures (see Section 5.4.2.1) that reduce the effective viscosity. Laboratory experiments conducted on dry granular mixtures show that high deformation rates are mainly developed within a boundary layer at the bed. The effect of fluidification could therefore be taken into consideration by means of a further boundary condition, which can be expressed with a Coulombic friction law, at the bed, with the angle of friction of the bed reduced [40]; however, according to Takahashi [21], in "saturated" debris flows, being subject to high deformation rates, the entire depth of the flow should be considered as a boundary layer.

3.4.4 Particle segregation

As previously observed, although density remains more or less constant, debris flows are composed of granules of different sizes and shapes. When this kind of material is deformed, by the action of a gravitational field, the characteristic phenomenon of particle segregation or grading takes place. During this process, granules of the same type (shape and size) are redistributed during movement and tend to group together in homogeneous classes. Therefore, as a consequence of grain-size grading within debris-flows, the mean dimension of the particles varies from the bottom to the top of the streamflow depth. Grading can be *direct,* or normal, when the coarser particles accumulate at the base and their dimensions gradually shrink towards the top of the flow; or *inverse* when the opposite occurs. Inverse grading is a distinctive characteristic of debris-flows, considering sufficiently long observation timeframes, it produces the accumulation of coarser clasts at the head and on the free surface and a considerable slowdown in such areas. From a dynamic point of view, this mechanism, which can be observed during deposit analysis, can be qualitatively described as follows: the segregation of coarse material to the free surface means that this material moves more rapidly than that in the layers below, which carries it forward towards the head of the flow.

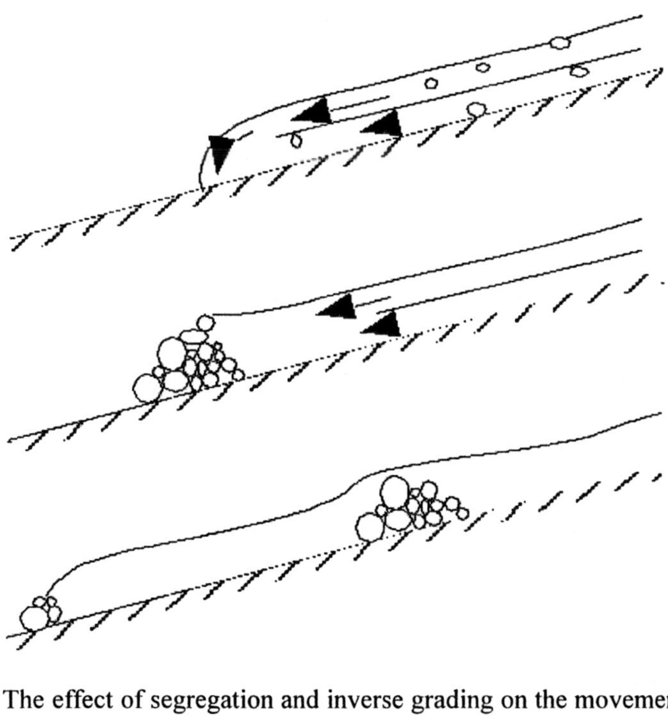

Figure 3.8: The effect of segregation and inverse grading on the movement of the head.

The head therefore proceeds in a "rotating" manner, sometimes compared to the movement of a conveyor belt or a caterpillar track, accumulating the larger boulders due mainly to an increase in friction in this area (in some cases, the "snout" of the debris flow can consolidate and create obstacles); the clasts later shrink once again into the flow and a new segregation cycle begins (Figure 3.8). A number of mechanisms have been proposed to explain this particular debris-flow phenomenon; Savage and Lun [35] propose an explanation based on *kinematic sieving*. If one considers adjacent layers of particles, parallel to the main direction of motion and in relative movement on top of one another, at any instant, it is possible that chaotic fluctuations cause particles from the upper layer to fall into the empty spaces between the particles of the layer below. Inverse segregation therefore takes place because the probability of finding an empty space suitable for a smaller particle is greater than the corresponding probability for a larger particle, in such a way that the smaller particles tend to deposit on the bed, whilst the larger ones migrate upwards. Another explanation can be put forward using the collisional dispersive model characterised by a velocity profile convex towards the top; by which the lower layers would be subject to a greater shear rate and characterised by more intense collisions, therefore, the granules in the flow would be struck more frequently from below than from above, thus producing an upward movement that would affect coarser granules in particular, given that dispersive stress is proportional to the square of the diameter. A realistic description of the phenomenon must therefore consider both particle-size changes, using an equation that describes the space–time evolution of the grain-size of the constituent material, and the angular momentum, which would constitute a far more complex structure than the models traditionally used for water–sediment mixtures.

3.4.5 The principle mechanisms of particle dispersion

The resistance to flow of a debris flow is represented by the sum of the resistance due to deformation and the mixing of the particles dispersed in the fluid and that due to the consumption of energy due to particle dispersion; the greater the concentration, the greater the resistance will be. The energy required by the dispersion mechanism comes from the action of stress; in equilibrium conditions, resistance to flow balances the active force of gravity. On gentle slopes, the force of dispersion becomes less intense due to the drop in velocity, the gradient itself diminishes and the particles tend to deposit. The dispersion mechanism of the particles in the flow is linked to the properties of the material, the angle of slope and concentration and also conditions the behaviour of the flow, which can be *stony*, *muddy* or *viscous* [41]. Many of the fundamental concepts used to study debris-flow dispersion are still linked to the dispersive theory proposed by Bagnold [7]. However, observations made on solid clasts suspended inside debris-flows in quasistatic conditions demonstrate that dispersive pressure alone is not adequate to completely describe the mechanism of a coarse clast suspension in water–sediment mixtures.

If a particle is suspended in a steady slurry, when subject to stress, it tends to deposit and remain in that position without further movements. This brought Johnson [9] to deduce that another suspension mechanism had to intervene in combination with dispersive force, a mechanism that he termed resistance of the interstitial fluid. Flocculation, which is caused by the action of forces of attraction between the clay particles, would be responsible for this resistance of the interstitial fluid, defined as the minimum stress required to break the bonds between the flocculated particles. The resistance of the interstitial fluid therefore seems to be connected to the presence of clay in the flow, which poses the problem of calculating for which percentage of clay this mechanism becomes significant for the suspension of clasts in a debris flow. It has also been observed that the resistance of the interstitial fluid tends to disappear for high deformation rates; in reality, according to Davies [42], when the flow is in static conditions, stress is transferred to the bed through the interstitial fluid, however, once the threshold condition has been exceeded and flow occurs, residual stress causes a gradual increase in the shear rate in the layer subject to continuous deformation, even when an undeformed rigid zone persists.

With regards to previous mechanisms of particle dispersion in a debris flow, the buoyant force is particularly evident; in a silt and clay suspension, clasts are dispersed throughout the flow and the buoyancy (Archimedes' force) is the weight of the displaced water. If stream capacity is given by the mean diameter of the largest particle that can be transported in the flow, in the same way, suspension capacity can be defined as the largest particle that can be kept in suspension within the flow. The pressure gradient that causes floating can be subdivided into two components, the first being that typical of Newtonian fluids, in other words, hydrostatic pressure, whilst the second represents the excess of intergranular pressure due to the transfer of part of the applied load, previously supported in friction contacts between grains, to the interstitial fluid, thus increasing pressure (see Terzaghi's principle of normal effective stress in Section 4.2.2). The excess of pressure is certainly important in reducing the resistance of the material during motion triggering, however, its contribution to the mobility in a system subject to continuous deformation appears limited [1]; in fact, in such conditions the interstitial pores are continuously destroyed and remade and it is therefore very easy for the fluid and the grains to dissipate any excess pressure.

3.5 Energy in a debris flow

The interactions that take place within the mixture make debris-flow energetics completely different from that of homogeneous solid or liquid substances. If we start with the assumption that a debris-flow consists of a collection of elastic, smooth and identical spheres positioned along an inclined rigid plane that ends with a horizontal deposition surface, immersed in a vacuum; obviously, as there is no energy dissipation during movement, the phenomenon can proceed infinitely. If the interstitial vacuums between spheres are then filled with a less dense liquid (for example water), due to viscous dissipations, sooner or later the mixture will stop, once it reaches the horizontal plane. If we then substitute these ideal spheres with inelastic, rough granules immersed in the vacuum, the mass will stop due to the friction in the contacts between granules and inelastic collisions; by adding the interstitial fluid, by analogy one might think that stoppage is further brought forward by the viscous dissipation of the fluid. However, experiments conducted on granular mixtures "saturated" with water, show that, despite the intervention of viscous dissipation, stoppage is delayed [3]. This abnormal behaviour is the result of interactions between the viscous fluid and the collection of inelastic granules of various sizes over a rather vast interval.

From the initial motion condition to deposition, within a debris flow a cascade of energy conversions takes place that leads to its degradation and to the consequential increase in entropy of the system (Figure 3.9).

The net efficiency of a debris flow is linked to the conversion of potential gravitational energy into the work performed during translation; the more efficient this conversion, the lower the degradation of energy, thus limiting the irreversible conversion into heat, the greater the distance L covered before stoppage (runout).

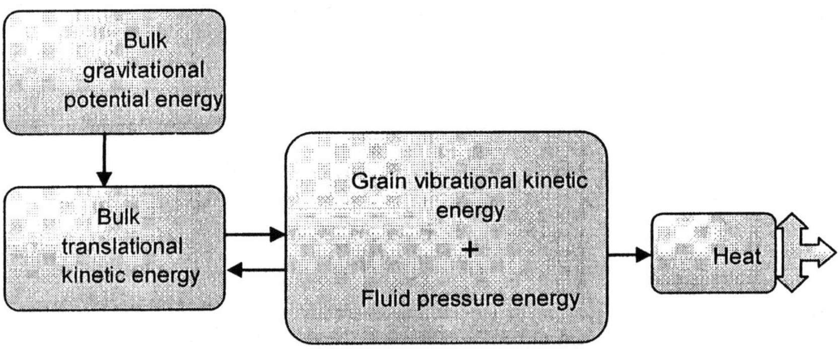

Figure 3.9: Energy conversions during a debris flow.

Net efficiency can be calculated by integrating the equation that describes the motion of the centre of the debris flow mass as a temporal function; alternatively, it is possible to extrapolate an estimate by equating the total potential energy MgH, relative to the height H in relation to the stoppage plane, to the energy dissipated over distance L, which is the work performed by the resistive forces MgR until stoppage [18]:

$$MgH = MgRL\,, \tag{3.7}$$

M is the overall mass of the debris flow, which is assumed an isothermal system, g is the value of gravitational acceleration and R is a dimensionless resistance coefficient that incorporates the effects of the internal forces, but that also depends on the external forces that act on the bed, which intervene in the conversion of the initial potential energy into kinetic energy; the spatial coordinates refer to the centre of mass, in such a way that H is the height of the initial pile of debris evaluated in relation to the final deposit, and L is the horizontal distance measured in a straight line. By dividing each member of eqn (3.7) by $MgHR$ one obtains the net efficiency, defined as $1/R$:

$$1/R = L/H\,, \tag{3.8}$$

it therefore increases with the stoppage distance for a fixed height.

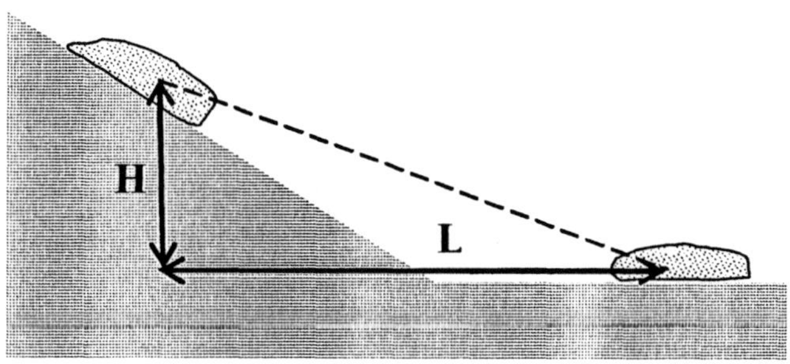

Figure 3.10: Coordinates of the centre-of-mass of the initial accumulation of debris and deposition.

This formulation of net efficiency as a function of L and H allows us to obtain estimates for it by means of recordings in the area interested by the debris flow. Some of these estimates show that [18]:

- the values of L/H, for "saturated" debris flows are greater than those for granular mixtures and comparable masses;
- large–scale debris flows have greater efficiencies than small ones;
- L/H depends closely on the geometry of the pathway covered and the conditions of the bed (presence of erosion or sedimentation).

It can be observed that for debris flows with volumes greater than 10^5 m^3, *L*/*H* increases more or less in proportion to the logarithm of the volume, however, for smaller debris-flows this value may remain almost fixed at values of between 2 and 4 [18].

In agreement with eqns (3.7) and (3.8) mass should not influence the efficiency parameter, however, data challenges this hypothesis; the cause of this discrepancy is difficult to pinpoint, as during motion, mass and composition can vary independently of one another. Variations in the mass of a debris flow are usually indicated using the terms *bulking* (increase in solid concentration) and *debulking* (reduction in solid concentration). The attempt to use an elementary balance to forecast the effects of variations in mass on efficiency is incompatible with the implicit assumptions of eqns (3.7) and (3.8); for example, it could be argued [18] that the reduction in mass during motion could increase efficiency, because the potential energy that is initially available remains fixed, whereas the work performed by the resistive forces drops with the reduction in mass. This would be correct if *R* depended on internal forces alone, however, it also depends on external forces that can contribute to mass loss. The reduction in mass requires additional work to be performed on the flow by the banks and channel bed in order to decelerate or deposit the lost mass; this work is added to that performed over the same distance in the absence of deposition. The unresolved problem is therefore to establish whether the additional work is less than the energy saved following mass loss and if an accumulation of mass would increase the stoppage distance and debris-flow mobility.

3.5.1 Granular temperature

The study of debris-flow dynamics has changed many concepts of the rheometric investigations performed on dry granular mixtures and solid–liquid mixtures [7, 17]. A granular mixture is defined as a collection of a large number of discrete solid particles, whose interstitial spaces are filled by a fluid (air, water). As is the case for debris-flows, granular mixtures are multiphase systems, however, one can talk for a granular mixture in stricter terms when it is assumed that the role of the interstitial fluid can be neglected in describing flow behaviour. By definition, in a granular mixture, the particles are strongly massed together and closely packed, or much denser than the fluid itself, in such a way that only the particles (and not the fluid or the particle–fluid interactions) are central to momentum exchange. The parallelism that exists between granular mixtures (in particular for high deformation rates) and debris-flows, has been successfully employed in describing the intergranular interactions and the semifluid behaviour, which are common characteristics between the two systems, of the mixture as a whole; it will be used here to illustrate a key concept for the comprehension of both granular mixtures and debris-flows.

In a rapid flow regime, the velocity of each particle can be broken down into the sum of the mean system velocity and an apparently random component, which describes the motion of each individual particle in relation to mean motion. The particles would therefore float in the interstitial fluid as atoms in an ideal gas; this physical analogy between the chaotic motion of granular particles and the motion of the molecules in the kinetic theory of gases, appears so explicit that we generally refer to the mean value of the square of the chaotic velocity of the particles, with the term *granular temperature* or simply *temperature,* an expression introduced for the first time in the study of granular mixtures conducted by Ogawa [43]:

$$T = \left\langle \left(u_p - \overline{u_p} \right)^2 \right\rangle \tag{3.9}$$

where u_p is the instantaneous velocity of the particles, $\overline{u_p}$ is the mean velocity of the particles at a certain point and their difference represents the instantaneous deviation from mean motion; T is therefore double the kinetic energy per unit of mass contained in the chaotic motion of the particles. Thermodynamic temperature and granular temperature would play similar roles in determining the behaviour of the respective systems; in particular they both generate pressures and govern the processes of internal transport of mass, momentum and energy. As macroscopic collisions between solid particles are inelastic, granular and molecular systems have significant differences, in particular, the energy associated with granular temperature is continuously dissipated by collisions between the particles; it therefore goes without saying that the granular temperature is gradually degraded. Shaking a granular mixture causes chaotic motion between particles, but if shaking is suspended, the particles almost instantly lose their granular temperature and stop moving; in order to supply granular temperature, energy would have to be continuously provided to the chaotic components of the velocity, in order to balance the amount subsequently dissipated by the collisions. Moreover, in contrast to molecular systems, granular temperature is a byproduct of the flow itself and it cannot therefore be specified irrespective of the mechanics of the flow, which is, however, possible for molecular systems, thus comparing it with, for example, a large tank of isothermal heat. There are two main transport mechanisms in a mixture [35]: transport by collisions, due to particle interaction, and the kinetic transport linked to particle fluctuation. In the former case, granular temperature constitutes a byproduct of intergranular collisions. In fact, when two particles collide, their resulting velocities do not depend on their intrinsic velocity alone, but also on the angle of collision, the friction surface at the point of contact and other factors that could influence the system's geometry. Therefore, although initially the particles move with suitable mean velocities for their position within the velocity field, their velocities caused following collision will have components of chaotic velocity. The chaotic velocities generated will thus be proportional to the velocity of each particle at the moment of impact, for which, in mean terms, they must also be proportional to the mean shear rate within the mixture. The second transport mechanism is associated with the disorderly motion of the particles between two collisions; on its course a particle, moving parallel to the local shear rate, will assume an apparently chaotic velocity that is approximately equal to the difference of the mean velocities corresponding to its current position and that of the last collision that took place [35]. Once again the chaotic velocity created in this way will be proportional to the local shear rate, however, in contrast to the previous mechanism, it can only generate a chaotic velocity component, in other words the component in a perpendicular direction to the mean shear rate, therefore, the temperature generated in this way will be anisotropic with a considerably higher component in the direction of mean motion (even if the subsequent collisions will contribute to deviating the direction of chaotic velocity) [35]. The first way of generating granular temperature will be prevalent in strongly closely packed mixtures, where the particles are not free to move following collisions, whereas the second way, which causes anisotropic temperatures, will dominate when the particles are very dispersed and collisions are infrequent. The granular-temperature concept is central to the behaviour of granular mixtures in rapid motion regimes and its understanding requires the analysis of the processes linked to the internal energy of the flow; to this end, Figure 3.11 shows an extension of Figure 3.9, in which the concept of granular temperature is introduced with reference to the mixture's internal processes.

The source of all the energy available in a granular system is the work performed on it from the outside, both by body forces (in the absence of electromagnetic forces the only body force considered is gravity) and surface forces (pressure forces and kinetic forces that therefore generate deformation). The shear work performed by the stresses (product of the shear stress and shear rate) on the system converts a part of the kinetic energy relative to the mean motion into granular temperature, in other words, into kinetic energy associated with the chaotic motion of the particles. This reflects the fact that both the granular-temperature generation mechanisms are correlated to the shear rate. Therefore, the inelastic collisions between particles will gradually dissipate the granular temperature, thus converting it into thermodynamic heat. The entity of the granular temperature will therefore depend on the compromise between the generation of temperature due to shear work and its dissipation by collisional effects.

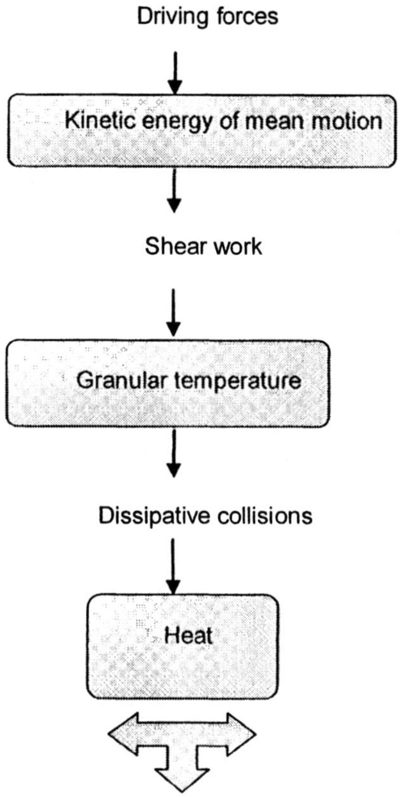

Figure 3.11: Mechanical-energy cascade of a granular mixture in a rapid motion regime.

This is the cascade of global energy in the mixture, however, locally the value of the granular temperature could be influenced by its entity in the surrounding areas, due to a process similar to the thermal conduction of a gas; when a particle follows its chaotic pathway within the system's overall motion (kinetic transport due to particle fluctuations) and when two particles collide, part of this energy is exchanged between them (energy transport by collision). Both mechanisms bring about a diffusion of granular temperature in the system, which, by following its gradients, takes place in the direction of the lower temperatures. One characteristic aspect of granular materials that substantially distinguishes them from that of gases, is that the chaotic velocities, being proportional to the mean shear rate, are much smaller than the mean velocities of the particles and, for this reason, the fluctuations of the particles are not so violent as might be expected.

3.6 Debris-flow propagation

3.6.1 Triggering and mobilisation

The most frequent cause that triggers debris flows is certainly the collapse of hill- or mountainsides. Alternatively, debris-flows may also occur due to the interference of a series of small landslides during descent. Triggering is caused when an initially static and apparently rigid debris pile reaches the instability condition; the subsequent mobilisation requires the simultaneous occurrence of three conditions [18]:

- collapse of the mass;
- its "saturation" with water;
- an adequate conversion of potential gravitational energy into internal kinetic energy, in order to allows the passage from a movement mode connected to the localised shearing on the collapse surface, to a more widely diffused deformation recognisable as flow.

In certain debris flows, the water required for the "saturation" of the material comes from outside following the collapse, however, in most cases, all the water required is within the mass when the collapse takes place. Triggering itself is often caused by a variation in the intergranular pressure that follows the infiltration of water due to superficial meteorological phenomena. In certain cases, the subsequent mobilisation is facilitated by the contraction of the debris mass, thus causing a transitory excess of pressure that facilitates the second and third conditions. As an alternative to contraction, mobilisation is possible if there is sufficient energy to shake and disaggregate the mass during the landslide, as well as the presence of water due to "saturation".

In the initial motion condition, the well-known Mohr–Coulomb criterion eqn (3.5), provides an adequate description of the state of stress acting on a surface, by means of the balance between mean active stress, the resistant stresses in the friction contacts between grains and cohesion c. During the collapse, the cohesive bonds break gradually and therefore at the limit $c \approx 0$ even in particularly clay-rich mixtures, therefore resistance to motion is mainly given by the angle of internal friction and effective normal stress, herein defined as the difference between total compression stress and the pressure of the intergranular fluid [18].

3.6.2 Developed motion

Following mobilisation, debris flows flow as a continuously shaken mass that resembles wet cement; the larger flows can transport boulders up to 10 m in diameter. The typical motion of these phenomena is characterised by one or more non steady and heterogeneous waves that occur at irregular intervals; the larger waves tend to engulf the smaller ones leading to the subsequent coalescence of the corresponding heads, which can produce a sequence of considerably sized, unstable waves (Figure 3.12). The mechanisms that control the motion of the debris-flow are the pressure in the fluid and the granular temperature, which act in synergy within the flow, as the energy cascade proposed in Figure 3.11 shows, in turn, the motion of the flow feeds both granular temperature, by means of the conversion of kinetic energy of translation of grain fluctuation and a non hydrostatic distribution of pressure. High pressures in the interstitial fluid reduce friction and influence the collisions associated to a high granular temperature; therefore, the understanding of the origin and the effects produced by high pressures in the fluid, is the key to debris-flow analysis. One possibility is that the rapid tendency to contract that some loose materials have, produces high intergranular pressures in debris-flows and therefore conditions similar to liquefaction (see Section 4.2.2).

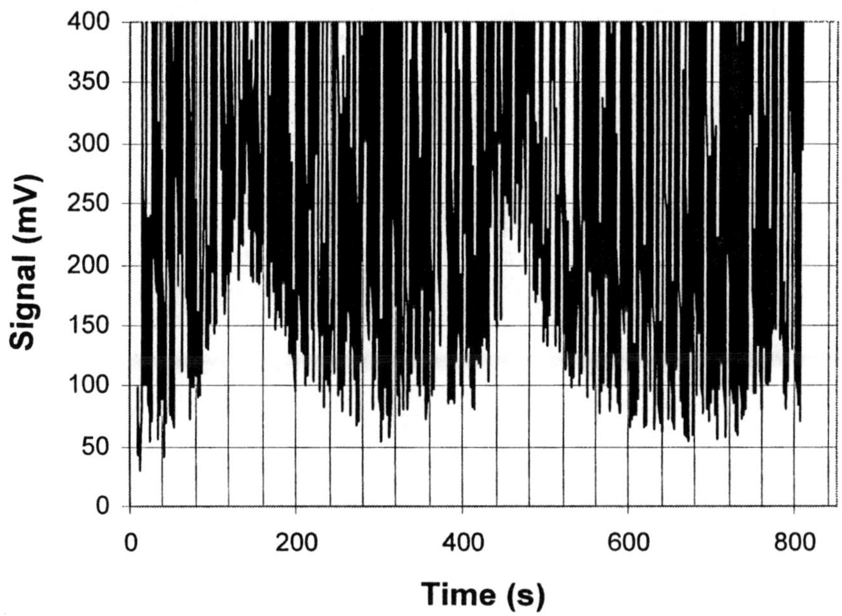

Figure 3.12: Registration of vibrations induced by the passage of a debris flow on 25th July 1998 at Acquabona (Bl, Italy) made using geophones; it is possible to distinguish the passage of the various flow waves [44].

This rupture-collapsing behaviour, particular to certain types of ground, however, requires the mass to contract progressively, a condition that cannot be supported by a steady debris-flow; on the other hand, a debris-flow in steady motion has not only never been observed in nature, but is also difficult to create using experimental apparatus. Therefore, much evidence supports the hypothesis that debris-flows are fundamentally nonsteady phenomena and rheometric experiments or theoretical models that start from the assumption of steady motion may provide a valid starting point for the analysis of such phenomena, but do not make it possible to shed full light on it.

3.6.3 Stoppage and deposition

From a geographical point of view, debris-flows form particularly characteristic deposits known as alluvial cones or fans, which are steeper than normal alluvial basins. Debris-flow deposits differ from those of a generic water streamflow with ordinary solid transport, especially with regards to particle sorting: whereas debris-flows have coarse, scarcely sorted granules immersed in a suspension of finer material and in some cases, isolated rocks of considerable dimensions can be encountered, the latter are mainly composed of uniformly sorted, fine particles. The presence in debris-flow deposits of elongated particles arranged in the direction of the flow has sometimes been considered as proof of the laminar nature of the phenomenon [1]. Stoppage of a debris-flow mainly takes place in two situations: along its course the flow encounters gentle slopes or areas less restricted than the channel of descent. In any case, an equilibrium condition between the tangential component of the weight and the forces of internal and external friction is once again reached. Takahashi [14] studied the deposition process of debris-flows in the case of a drastic reduction of the slope angle with an equal channel width; considering the variation in momentum at the head, and moving closer to the area of gentle slope, he obtains the condition for the stoppage of the head as:

$$\tan \vartheta_{\mathrm{f}} < \frac{C_{\mathrm{vm}}(\rho_{\mathrm{s}} - \rho_{\mathrm{l}})\tan \phi_{\mathrm{d}}}{C_{\mathrm{vm}}(\rho_{\mathrm{s}} - \rho_{\mathrm{l}}) + \rho_{\mathrm{l}}}, \tag{3.10}$$

in which ρ_{l} and ρ_{s} are the density of the liquid and solid, C_{vm} is the volumetric solid concentration uphill from the stoppage area, ϕ_{d} is the angle of dynamic friction and ϑ_{f} is the inclination of the slope in the stoppage area. The condition shows that for high interstitial fluid densities and low solid concentrations of the mixture, stoppage is possible along increasingly gentle slopes. Before stoppage, it is hypothesised that the flow flows in steady conditions along a slope with a constant inclination ϑ_{m} with volumetric solid concentration C_{vm}; this concentration can be obtained from the flow-trigger condition (see the analysis of stability proposed in Section 4.3.2, by substituting the generic ϑ and C_{v} by the current ϑ_{m} and C_{vm}), shown below:

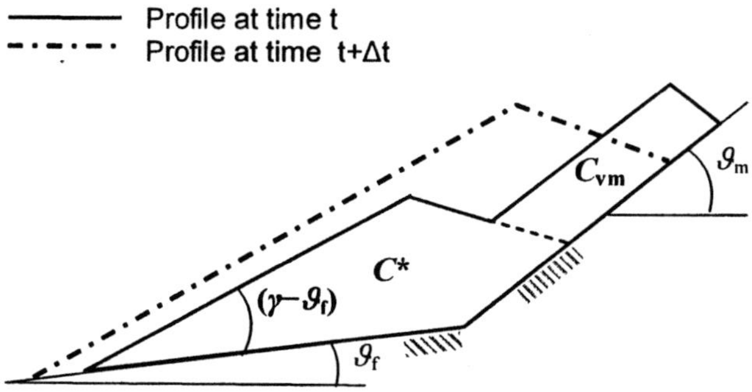

Figure 3.13: Diagram representing the deposition process [14]; C_{vm} and ϑ_m are the volumetric solid concentration and the inclination of the slope uphill from the stoppage area, ϑ_f is the inclination of the slope in the stoppage area, C^* is the maximum value of volumetric solid concentration and γ is the inclination of the stable pile of deposits.

$$\tan\vartheta_m = \left[\frac{C_{vm}(\rho_s-\rho_l)}{C_{vm}(\rho_s-\rho_l)+\rho_l}\right]\cdot\tan\phi, \tag{3.11}$$

in which ϕ is the angle of static friction. By inverting the previous relationship, the following expression is obtained for C_{vm}:

$$C_{vm} = \frac{\tan\vartheta_m\cdot\rho_l}{(\tan\phi-\tan\vartheta_m)\cdot(\rho_s-\rho_l)} \tag{3.12}$$

which, when substituted into eqn (3.10), provides the definitive condition for stoppage of the head:

$$\tan\vartheta_f < \frac{\tan\vartheta_m\cdot\tan\phi_d}{\tan\phi}. \tag{3.13}$$

As the angle of dynamic friction ϕ_d is lower than the corresponding static angle ϕ, the flow could reach slopes that are gently inclined despite the fact that it gradually slows down. Moreover, when the flow stops, the volumetric solid concentration reaches its maximum value C^*, this means that the interstitial fluid is gradually drained, flowing away over the accumulation surface. If the inclination of the deposits, immediately prior to stoppage, is higher at the critical angle for triggering mass wasting, the granules on the surface are in a situation of instability and continue to flow until a setup corresponding to the critical angle is reached; therefore an accumulation of deposits is stable if it has at least the following inclination (in order to obtain eqn (3.14), see Section 4.3.2):

$$\tan\gamma = \left[\frac{C^*(\rho_s - \rho_l)}{C^*(\rho_s - \rho_l) + \rho_l(1 + h_0 d^{-1})}\right] \cdot \tan\phi\,, \tag{3.14}$$

in which $\tan\gamma$ is the slope of the stable deposit in relation to a horizontal plane, h_0 is the height of the superficial flow and d is the mean diameter of the sediments. If the debris-flow head stops without causing a significant widening of the flow, the terminal portion assumes a lobed form as if the flow had suddenly petrified [14]; the theoretical profile of this terminal portion of the flow was studied by Johnson [9], especially in relation to the characteristic curve of the head. If we assume the mixture as a simple plastic substance with density ρ_m, yield stress τ_y and the surface in equilibrium in each point, the profile proposed for the deposits is as follows:

$$x = -\frac{\tau_y}{\rho_m} \ln\left[\cos\left(\frac{\rho_m}{2\tau_y} \cdot y\right)\right]. \tag{3.15}$$

Stoppage therefore causes the deposition of the mixture's solid fraction, however, this process again differs from the corresponding one in ordinary solid transport, in which the deposition takes place due to the dampening of the turbulence that suspends particles; in fact, a debris-flow is a heterogeneous entity and in the final phase the entire flow shows a progressive tendency to slow down. The stoppage process constitutes a special case of unsteady motion that takes place when all the kinetic energy still available degrades within irreversible forms. This process initially concerns the coarser granular portion of the flow, which is found in the head and on the margins (see inverse-grading phenomenon, Section 3.4.4) and in these areas, causes the cancellation of the granular temperature with the consequent formation of rigid banks that form the perimeter of what will be the main debris-flow deposit. The first deposits are therefore composed of the coarser material and generally form barriers that hinder or stop the motion of the fine fraction that arrives from behind and that still has high interstitial pressures. If more waves occur, the deposits accumulate on top of one another and in many cases can also be remobilised by the passage of a particularly destabilising wave. The complex interaction between the resistance of the first deposits and the momentum of the flow that follows produces accumulations that are solid and drained on the perimeter but diluted and not very resistant inside, in such a way that here the intergranular pressure can remain beyond the hydrostatic level, thus keeping the sediments in a practically liquefied state for a long time, even after deposition. The last step towards the transition of a debris-flow from a liquid state to a solid state is the subsequent decline in the pressure of the fluid, and relative solidification of the granular mixture (mainly due to sedimentation under the action of gravity) now completely devoid of the liquid part, following drainage and evaporation. Iverson [18] proposes a law to describe the decline in the intergranular pressure in debris-flow deposits, according to the square of a characteristic length chosen for the drainage process and a diffusive coefficient $D=kB/\mu$, in which k is the hydraulic permeability (see section 3.3.4), μ is the viscosity of the fluid and B (Pa) is composite stiffness. All materials have a volume elasticity and the variation in volume $\Delta V/V$ is proportional to the variation in the stress $\Delta\sigma$ that caused it, $\Delta\sigma = -B(\Delta V/V)$; given that an increase in pressure usually brings about a reduction in volume, the minus sign makes it possibile to write B, composite stiffness, as a constant of positive proportionality. Thanks to a number of observations made by Iverson [18] on typical deposits, in which length and width vastly exceed thickness, indicated here by h, it has been observed that drainage takes place preferably in a vertical direction; this led Iverson [18] to choose the thickness h as the characteristic length of the drainage process, and finally to formulate the following law for the decline of intergranular pressure:

$$t_{\text{diff}} = {h^2 \mu}/{kB} \tag{3.16}$$

in which t_{diff} is the time required for the decline of intergranular pressure.

3.7 The problem of saturation in debris flows

As mentioned in Section 3.1, a debris flow takes place when a mass of scarcely sorted sediments that is shaken and "saturated" with water, flows along a slope under gravitational force. In particular, the mobilisation process, which concerns an initially static and apparently rigid stack of debris, requires:

- the reaching of the mechanical conditions of instability and the consequential collapse;
- a sufficient quantity of water for the "saturation" of the pile;
- sufficient conversion of potential gravitational energy into kinetic energy.

The occurrence of all three conditions causes the transformation of a movement of prevalent shearing on a localised surface of a superficial landslide, in a motion that involves more general deformation and that can be commonly recognised as flow. In this sense, even a landslide, which slides along a steep slope, can evolve into a debris flow if, during its motion, sufficient water is made available for the "saturation" of the mass. Therefore, one salient aspect in the occurrence of a debris-flow is linked to the gradual transformation from solid mass to fluid mass due to the progressive incorporation of water; in fact, a debris-flow is often defined as a mixture of sediment "saturated" with water, without specifying exactly what is intended by "saturation of the mixture". In reality, the reaching of the yield condition can take place prior to complete saturation of the material, intended as it will be defined later; therefore, the deformation of the sliding mass gradually allows water to mix with the debris, thus reducing resistance and causing the consequent liquefaction due to exceeding the hydrostatic value by the pressure in the pores, until effective stresses are annulled. Therefore, in a material in a fluid state (a mixture of granules and fluid) that descends along a channel, a situation very similar to saturation can take place, in which there are negative pressures generated by capillarity, as well as positive neutral pressures (or null). Although this book does not directly deal with the subject of the specific dynamics of the sub-superficial flow, and therefore connected to any filtration phenomena or those of gradual water-table growth, or contraction, that some soils can show, the dynamics of which are dealt with in soil-mechanics studies, given the importance of the fluid phase and intergranular pressures in debris-flow studies, it would seem fitting to investigate closer the conditions that cause the filling of the pores in the ground. We will therefore abandon the macroscopic point of view and continue with analysis on an atomic-molecular level.

According to its physical status, three phases are present in soil: solid, liquid (often incorrectly termed interstitial water) and gas (or earth air). This distinction is made by grouping distinct phases together to form a single portion of material; this is particularly important for the solid part, which is usually very heterogeneous (for example, the solid constituents of a debris flow), and should, strictly speaking, always be considered polyphasal. The relationship between the various phases is strictly dependent on the grain-size of the medium: the size of the pores (which conditions the circulation of air and water) is, in fact, a direct function of the dimension of the various solid particles and their mutual spatial arrangement.

3.7.1 Introductory definitions

When we consider the movements of water within the soil, an important hypothesis is usually introduced; although the particles that make up the soil are subject to relative movements between them, however varied, the soil is assumed to be a porous, rigid medium, within which the particles are arranged rigidly with regard to one another. In reference to a soil sample, below are shown the main definitions that can be useful in characterising the movements of water within it:

- the *apparent volume* V_A of a soil sample is defined as the volume that the entire sample occupies, including the pores full of air and/or water inside it;
- the a*pparent density* is defined by the relationship $\rho_A = \dfrac{m_s}{V_A}$ between the mass of the solid part and the apparent volume of the soil; therefore, this definition excludes the mass of the water, and that of air is considered negligible;

- *real density* is defined by the relationship $\rho_S = \frac{m_S}{V_S}$ between the mass of the solid part and the volume occupied by the solid part alone; it therefore represents the mean density of the solid part;
- *porosity* is defined by the relationship $P = \frac{V_a + V_l}{V_A} = 1 - \frac{\rho_A}{\rho_S}$ between the volume occupied by the liquid part and the gaseous part and the apparent volume of the soil. Porosity is a property of the soil that can vary with time and often has a value of between 0.4 and 0.6 for nonclayey soils and reaches higher values for clayey ones;
- the *pore index* is defined by the relationship $e = \frac{V_a + V_l}{V_S} = \frac{P}{1-P}$ between the volume occupied by the liquid and gaseous part and the volume of the solid part;
- *ground humidity* is defined by the relationship $U = \frac{m_l}{m_s}$ between the mass of the liquid part and that of the solid part;
- the *volume fraction* of the water is defined by the relationship $\theta = \frac{V_l}{V_A} = U \cdot \frac{\rho_A}{\rho_l}$; soil humidity can also be expressed by the *humidity ratio* $\vartheta = \frac{V_l}{V_S} = \frac{\theta}{1-P}$. For soils "saturated of water" one has $\vartheta_{sat} = P$ and $\vartheta_{sat} = e$, which means that the total volume of the pores and the volume of water coincide, or rather, all the pores are full of water;
- *the degree of saturation* is defined by the relationship $S = \frac{V_l}{V_a + V_l} = \frac{\theta}{P}$ between the volume occupied by the liquid part and that occupied by the liquid part and the gaseous part. This relationship is 1 for ϑ=*P* in other words, for "water-saturated" soils.

The water in a soil sample is not only *contained* (concept of humidity), but is also *withheld* (implication of force); therefore, work must be performed on the system to extract it, in order to provide sufficient energy to overcome that with which the water is held inside. The minimum work required per unit of volume, that the forces applied from outside on the soil sample considered conduct in order to isothermically carry an infinitesimal quantity of water from a standard reference position, assumed at zero potential, to a particular point in the system considered, is termed the *volumic potential of water;* it therefore represents the value of the energy with which the unit volume of water is held in the soil. It is a negative quantity in that it is work performed on the system, we also talk of a minimal quantity because the corresponding quantity of energy can increase significantly due to the additional presence of resistances inside the liquid in motion; in particular, the minimum work condition is that which corresponds to a reversible process. Being work to exert, potential can be expressed in dyne/cm^3, atmospheres or bar (1 bar=10^6 dyne/cm^2, 1 bar=0.987 atm, 1 bar= 0.1 MPa); it therefore has the same dimensions as a pressure and in this way it is equal to the negative value of the relative pressure of the water withheld in the soil, in other words, referred to the sum of atmospheric pressure and the internal pressure of the liquid. It may be observed that, as motion always takes place from regions of higher pressure towards regions of lower pressures, the water contained in a soil sample does not run out due to the action of negative relative pressures.

3.7.2 The state of water in the soil

The fixture and movements of the water contained in a soil depend on the type of forces that are exerted on the volume of water within a given soil. The sum of the forces acting on the soil water, all of which can be expressed as potentials, in other words as force fields that act on the same object, and therefore the state of water in the soil, can be summarised in the following formula:

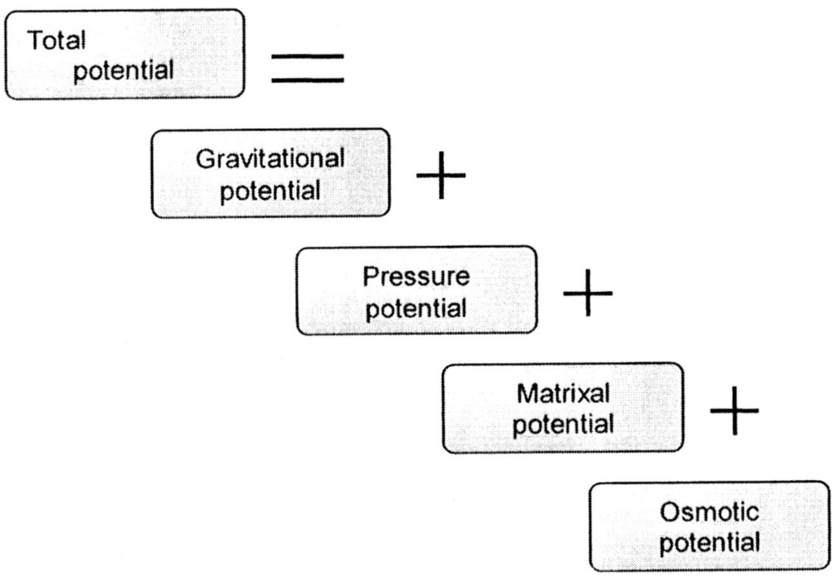

Figure 3.14: Total potential of the soil water.

- Gravitational potential: as any body, water is subject to the action of gravity; if free, under this force water will tend to move downwards (fall, or in the case of soil, percolation). Gravitational force is independent of the type of soil and is always the same, proportional to the constant g, for a given water present at the same depth of the soil.

- Pressure potential (hydrostatic pressure): is the sum of atmospheric pressure and the pressure exerted by the mass of the water in the soil above; once again, this force also tends to push downwards.
- Matrixal potential: water potential, usually negative for solids that can be wet by the liquid, which can be attributed to the effect of mutual attraction between molecules of the liquid and molecules of the solid matrix and therefore to the phenomena of capillarity (liquid in depression away from solid walls) and of superficial adsorption (films under pressure near the walls), which can coexist in different points of the soil with equal matrixal potential.

The force of capillarity is the consequence of the superficial stress of water and the adhesion of its molecules to the surface of the soil particles that "delimit" and "outline" the free spaces within the soil, and that is, in the end, to the walls of the pores. Surface stresses are similar to those created within a capillary following the development of two menisci, an internal one and an external one; the visible effect of these forces is the rising (if vertical) or the advancing (if horizontal) of the water contained.

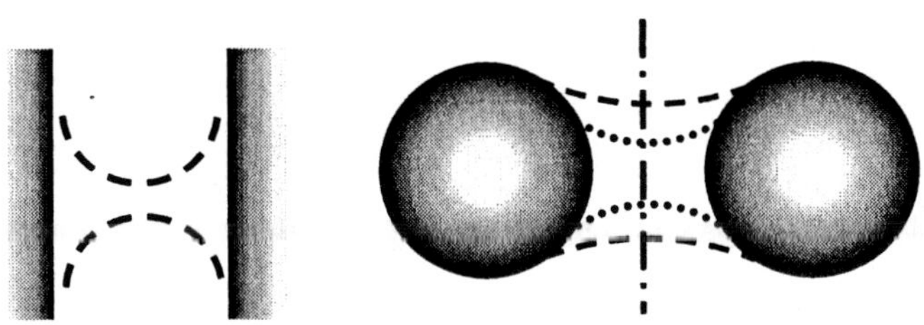

Figure 3.15: State of the water a) delimited at both ends by the water–air menisci of a capillary; b) compressed between two ideal spheres. In this case, it can be observed how, after partial asportation, the water that remains occupies a surface with a smaller curvature (dotted line) than the initial one (dashed line).

In the case of concave menisci, the water below the water–air surface is subject to a relative negative pressure caused by the forces exerted by the meniscus (which tend to push the water upwards) and by the weight of the water that adheres to it by cohesion. Therefore, in order to remove the water from the pore, work must be performed, per unitary quantity of liquid, in order to overcome the negative pressure of capillary potential; this is directly proportional to the surface stress of the water and inversely proportional to the radius of the meniscus and therefore, the lower potential (which is a negative quantity), the greater the work that is required and the greater the stress of the water to overcome. In the case of materials that are not perfectly wettable, the angle of contact of the air–liquid–solid system is greater than zero, thus causing an increase in the radius of the meniscus and therefore the drop in water stress. If we take the water compressed between two ideal spheres, a state of stress still exists; however, the extraction of the water contained not only causes a variation in the overall volume of water, but also an alteration in the curve of the surfaces (the mean radius is reduced and therefore the stress increases), which causes an increase in the potential of the water in absolute terms.

The adsorption force is responsible for the phenomenon by which a chemical species tends to concentrate on the surface of another, sometimes with the penetration of one into the lattice of the other and the formation of a superficial solid solution of the former on the latter; this phenomenon is attributed to the action of surface forces (the forces of Van der Waals, hydrogen bridges, and Coulombic forces). This phenomenon corresponds to the formation of a very thin film of water around the solid particles, with the possibility of penetration into the intermolecular spaces; the water molecules are under pressure and withheld by the attraction forces exerted by the solid wall and that are all the more intense the nearer the molecules are to the wall. For equilibrium reasons, the thickness of these films is lower the greater the stress of the water in the pores, or rather, the lower the water potential.

- Pneumatic potential is relative to the pressure (usually positive) of the gaseous phase on the water–air surfaces within the soil and is entirely transmitted to the liquid phase by combining with other water potentials. For the air pores of the soil that border with the exterior, this potential is clearly null (at the reference value), for occluded pores, the trapped air exerts a higher than atmospheric pressure on the aqueous menisci.
- Osmotic potential is caused by the concentration of solutes dissolved in the water and represents the work required to separate the water from the solution leaving the solutes within (for example, by evaporation) and thus overcoming the forces of attraction between molecules of water and solute. This contribution is due to the presence of ions in the circulating solution (and as such originally dissolved in the water or entered into solution from the soil as it passes), which increase the potential energy of the water in relation to the reference state (pure water); its value is directly proportional to the level of ions in circulation and inversely proportional to the water level (quantity of water in the soil).

The way in which these forces, which condition the permanence of the water in the soil, and therefore the entity of the potential associated to the water present in that soil can be expressed, are strongly conditioned by physical properties such as grain-size composition and structure. Soils with sandy, silty and clayey grain-sizes, respectively, will progressively have a prevalent presence of microgranules and consequentially of micropores. It therefore follows that the increase in water withheld by the forces of surface stress is decidedly higher in a silty-clayey soil, than in a silty-sandy soil where macroaggregates prevail and therefore macropores, or the aliquot of gravitational water.

The *liquid-retention curve* represents the matrixal potential trend in accordance with humidity and, for a solid material that is prevalently wettable by the liquid, it expresses the empirical relationship to be experimentally calculated for each porous rigid body, according to which the potential decreases with an increase in liquid content (Figure 3.16). With equal materials, the retention curve depends on its structural state and on temperature, moreover, at low liquid content values, it is strongly influenced by the grain-size composition and by the quality of the surfaces, for example, with equal liquids, the potential is lower the greater the adsorption surface, in other words, in clayey soils, and is higher in sandy soils. At higher liquid contents, the value of the potential depends to a great extent on the soil structure and therefore on the frequency distribution of the soil pores, in particular, at equal and high humidity, potential is higher in soils with larger pores (coarse sands).

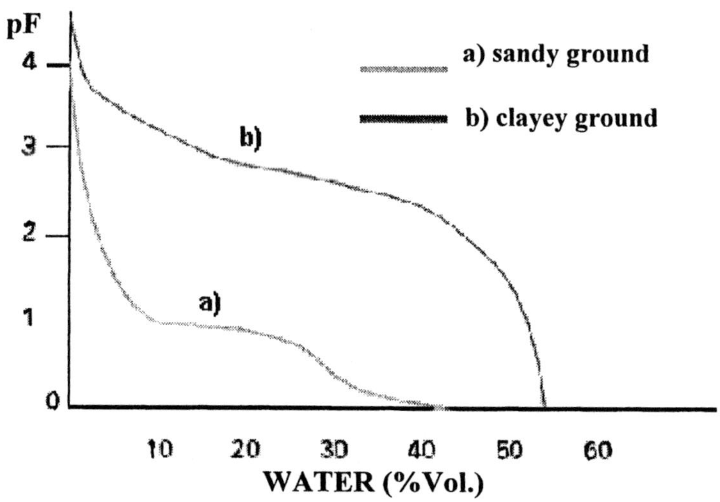

Figure 3.16: Retention curve. The relationship between the matrixal potential of the water in the ground (pF) and its relative content (humidity) in two types of ground. The pF (free energy potential) is the logarithm to the base 10 of the potential values referring to the unity of weight and expressed in cm of water.

The presence of the hysteresis phenomenon must also be considered: during the ground-drying phase, the water it contains is subject to a greater force than expected and, at equal humidity, higher than the force applied to it during the wetting phase; on the one hand this means that the ground withholds more water when drying but, on the other, that this water is withheld with greater energy and as a consequence, the greater is is the force required to extract it (Figure 3.17). This phenomenon can be, in part, attributed to the fact that the smaller pores, which clearly exert greater forces on the water than larger pores, slow down or prevent the runoff of the latter: if the water content in the larger pores must pass through small pores in order to runout, it will not be able to if they are always saturated with water at a pressure greater than that exerted by the water in the large pores; vice versa, during progressive wetting caused by rainfall or irrigation, the small pores once again slow down run off, however, in this case the water (in all classes of pores) is withheld with relatively lower forces than in the first case and hydrostatic pressure (in other words the weight of the water above, "waiting" to runoff) succeeds in making its action felt.

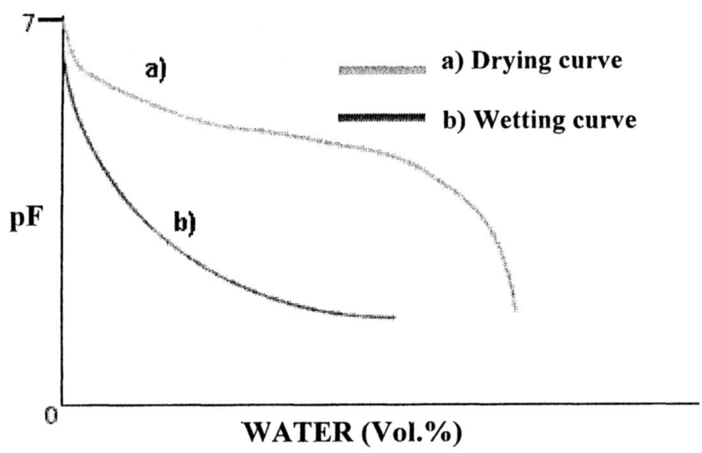

Figure 3.17: Wetting and drying curves: the hysteresis phenomenon.

Therefore, the retention curve associates a precise and calculated water content to each pressure value, according to the type of ground. In any type of ground, the maximum water that can be contained corresponds to a situation of total water saturation of all the pores; this situation corresponds to a value (percentage or absolute) termed *maximum capacity* and is linked to the ground's total porosity.

The initial hypothesis of the ground as a porous rigid medium excludes the possibility of the deformation of the ground's solid phase, however, in a non rigid porous material, as ground is in reality, the constituent particles can move in relation to one another, by means of translations, rotations, flexions and torsions, or combinations of these kinds of motions. If we consider two soil particles, assuming here the soil is a nonrigid porous medium, as two perfectly wettable spheres and neglecting the thickness of the film that entirely covers the two particles, we can highlight the relationships that exist between the deformability of the solid phase and the potential of the water. In fact, one consequence of the negative pressure that exists in the liquid included between the soil particles is the manifestation of forces directed along the join of the centres of the two spheres, thus tending to bring them together. By assuming constant the volumic potential of the water between the two spheres, it can be demonstrated (for demonstration see [45]) that the force exerted between the particles is proportional to the potential of the water and the area of the projection of the solid–liquid surface onto a perpendicular plane to the join between the centres of the two spheres. In addition, the irregularities of the shape of the particles under the action of the gravitational field can oblige them to perform rotational movements that can alter the spatial distribution of the particles themselves inside the ground. This spontaneously irreversible deformation can take place due to the effect of the phenomenon of capillarity when ground wetting takes place, or alternatively due to external mechanical actions or combinations of the two (the effect of driving rain). These alterations in the solid part of the ground and, in particular, its porosity causes a change in the ground's water-retention curve. Such variations have a considerable effect on the part of the water retention curve that corresponds to the higher water potential values (in modulus the lower values, those closer to zero, as we are considering negative values), therefore corresponding to higher water content and therefore larger pores.

Chapter 4

The state of the art of modelling debris-flow triggering and mobilisation mechanisms

4.1 Introduction

In recent decades a number of regions worldwide have been affected by debris flows. This has led to a parallel increase in interest from engineers in understanding how a "saturated" mixture of mud, rocks and organic debris can reach such significant levels of mobility and cover considerable distances on a wide variety of slope types. From the outset, research has always concentrated on understanding the physical mechanisms that govern the triggering of debris-flows along particularly steep slopes, their mobilisation and their relatively abrupt stoppage in the fan areas. The matter is complex and must be tackled from various points of view; sources of continuous interest are the dynamic behaviour of these particularly concentrated mixtures and, given their non–Newtonian nature, the constitutive laws that govern them. Since Bagnold presented his dilatant fluid model in 1954, a large number of other constitutive laws [7, 17, 46] have been put forward. One of the main drawbacks of these studies is linked to the verification, both in the field and in the laboratory; in the first case, in fact, the studies are often qualitative as they refer to mean flow parameters; in the latter they often suffer a number of restrictive factors. This chapter presents the state of the art in the literature of modelling debris-flow triggering and mobilisation mechanisms, in an attempt to identify those aspects that are as yet not fully understood and that inevitably condition the results that can be obtained from the application of mathematical models.

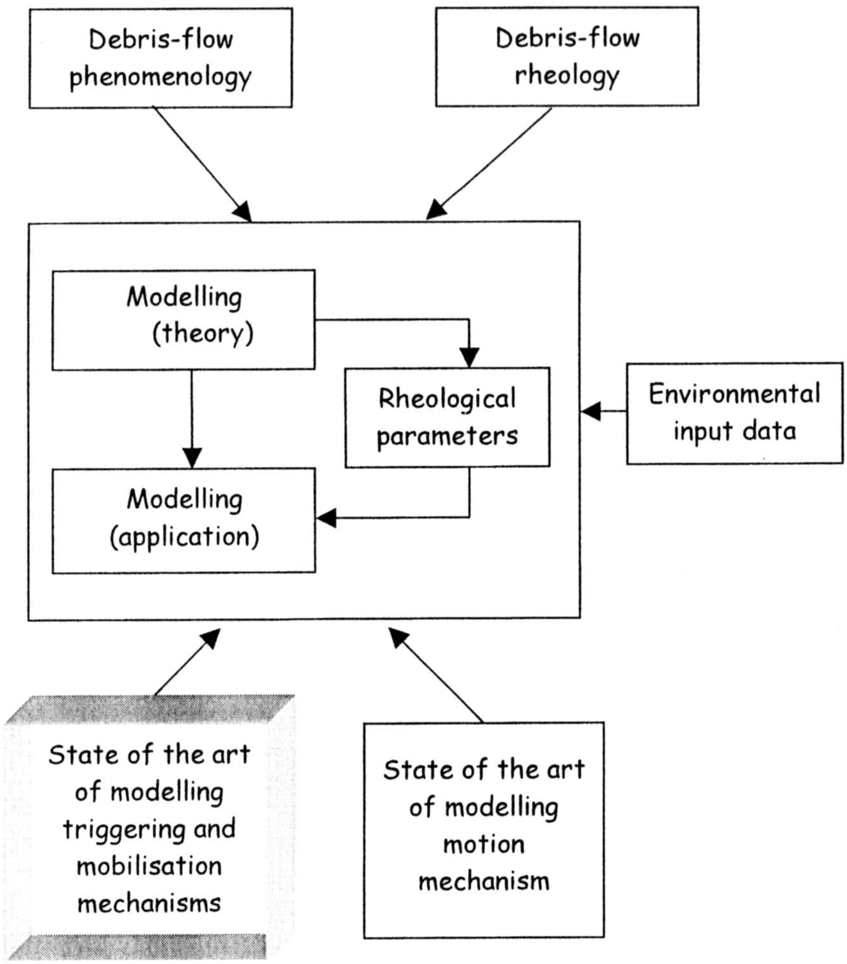

Figure 4.1: In the chart, showing a breakdown of the study approach adopted in this book, the grey frame highlights the analysis field of the state of the art of modelling triggering and mobilisation mechanisms, the topic that will be dealt with in this chapter.

4.2 Debris-flow triggering and mobilisation mechanisms

Debris flows are water–sediment mixtures that flow downhill under gravitational force. Given that the solid fraction is made up of a mass that has accumulated steadily along a surface, in order to have triggered a debris-flow, a force must have altered the equilibrium condition, thus causing instability for the most superficial portion of the accumulation at least, thus causing a generalised instability, as the balance of driving and resistive forces is in the favour of the former, and consequentially the dispersion of the particles inside a continuous flow. The possibility that this occurs can be associated mainly to two dominant processes [21]: a progressive increase in the flow's solid concentration and a change in the state (liquefaction) of the ground.

4.2.1 Progressive increase in the solid concentration of the surface flow

A mobilised debris accumulation does not initially contain a sufficient quantity of water to trigger a progressive "saturation" of the interstitial pores; therefore, in order to trigger a debris-flow, the appearance and sudden increase in a surface water flow is required. This first trigger mechanism presents a circular supply process that tends to increase the solid concentration of the surface flow, which occurs due to the subsequent instabilities induced by the passage of water over the bed material, until typical debris-flow values are reached. The quantity of water required for the potential triggering of a debris-flow may become available in the following ways: appearance of a surface flow following heavy precipitation, fluidification of a landslide and failure of a natural dam.

4.2.1.1 The appearance of a surface flow following heavy precipitation

Following heavy rain, the appearance of a surface water flow along steep slopes causes seepage and consequentially the destabilisation of accumulations of material that is then withheld, thus creating a flow. This case can be analysed using the models proposed for stream processes connected to bed-load transport (see Section 1.3.1); however, there are certain differences in the resistance law and in the erosion velocity equation.

4.2.1.2 Fluidification of a landslide

Debris flows can be triggered by the disaggregation of a mass shearing over a surface and the rapid mixing of debris with the water originally withheld in the mass itself or supplied during the phenomenon. Analysis can be performed by assuming a parallelepiped portion of the sliding mass with a lower part "saturated" with water; this part is subject to liquefaction due to the friction on the bed, and it therefore acts as a lubricant that facilitates the motion of the upper part. Progressive liquefaction causes the subsequent erosion of the mass above and therefore the static contacts between particles are broken, leading to overall "saturation" when the mass is entirely transformed into a debris-flow. The equations that describe the process show good correlation with experiments conducted on hydraulic channels and aimed at reproducing the phenomenon [41].

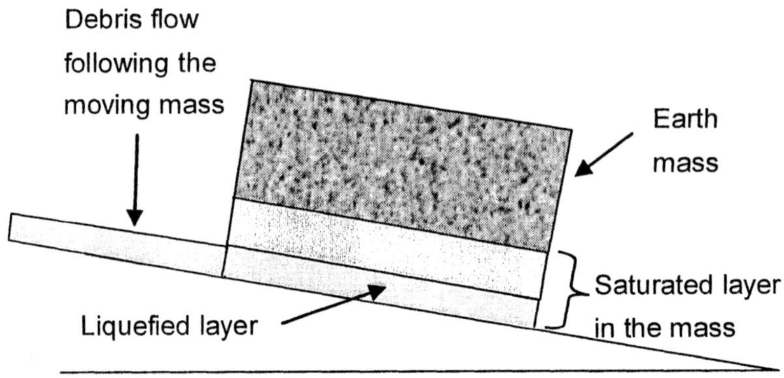

Figure 4.2: A mass moving along a slope with a partially liquefied layer [41].

4.2.1.3 Failure of a natural dam

In some cases, a landslide may obstruct a river; however, sooner or later the natural dam will fail, either due to the overtopping of the river or to the failure of the dam under the effect of seepage and hydraulic pressure.

In the event of *overtopping,* which is characterised by:

- stable dam;
- low permeability;
- considerable inflow discharge from uphill;

the dam resists seepage, until overtopping occurs. At this point, the outflow discharge causes the progressive erosion and imbibition of the downhill scarp.

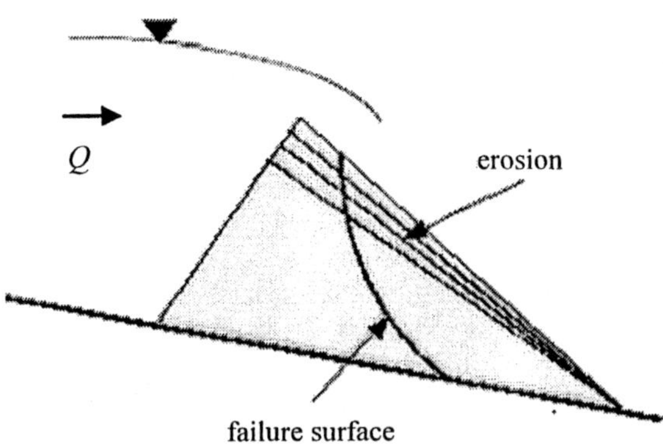

Figure 4.3: Debris flow following the failure of a dam caused by overtopping.

If the scarp is sufficiently steep, instability conditions may arise until a block eventually fails; as a considerable quantity of liquid is available, it can easily reach "saturation" and therefore produce a debris flow. This is often followed by the failure of the upper part of the dam resulting in a wave that amplifies the phenomenon.

The case of *sudden falls* takes place in the presence of:

- unstable dam;
- intermediate permeability.

Whilst the level uphill grows, seepage continues in the dam until the thrust of the basin uphill causes the upper part of the dam, weakened by the seepage, to collapse. As in the previous case, the fallen part is not "saturated", but may rapidly become so due to the liquid wave that forms.

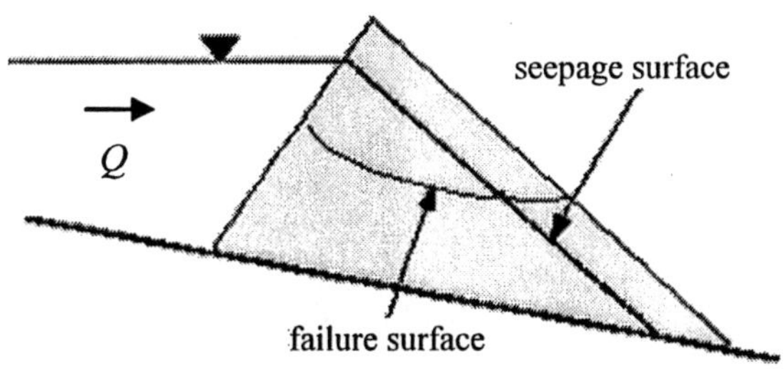

Figure 4.4: Debris flow following the failure of a dam caused by a landslide into a channel bed: sudden fall.

However, in the case of *a progressive fall,* which is characterised by:

- unstable dam;
- high permeability;

seepage rapidly reaches the downhill scarp, which is therefore progressively eroded from below, whilst the water level rises uphill. The combination of the two effects can lead to the failure of the upper part of the dam (not saturated) and the phenomenon proceeds as described for the cases above.

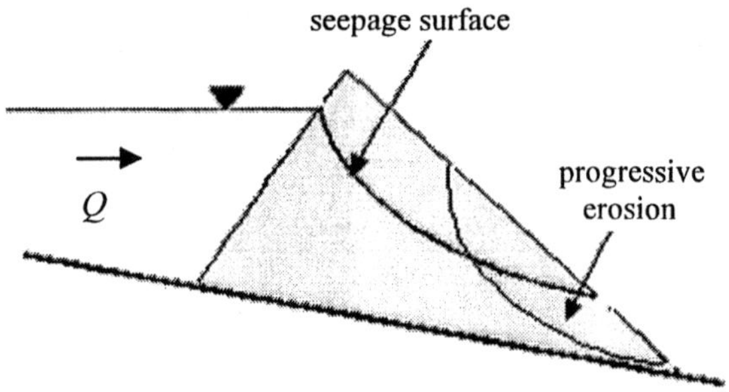

Figure 4.5: Debris flow following the failure of a dam caused by a landslide into a channel bed: progressive fall.

4.2.2 Change of state (liquefaction) of the soil

In the course of this second mobilisation mechanism, the mass of debris already contains a sufficient quantity of water to make it flow, and it is the initial structure of the soil (state of thickening) that is decisive in causing mobilisation. There are two types of behaviour that are dominant in the state of thickening of a soil with a high degree of saturation subject to shear strain: dilatant and contracting behaviour. In nondraining conditions, some "loose" soils, which therefore have a state of thickening lower than the critical density that tends to be followed by failure, may show contracting behaviour, in the sense that they tend to diminish in volume when subject to shear strain. In fact, the granules tend to rearrange in a more compact configuration, thus reducing the interstitial spaces and forcing the water it contains. These changes cause a sudden increase in the interstitial pressures. When a particularly compact soil in nondraining conditions is subject to the same type of deformation, the structure first undergoes a contraction, but subsequently, at greater deformation, it undergoes a progressive dilation. In these soils, the tendency to dilation and therefore the increase in volume brings about a decrease in the interstitial pressures.

Liquefaction is the process by which the internal friction resistance and the cohesion of a debris accumulation at high degrees of saturation are progressively reduced, following an increase in interstitial pressures, which develop under the application of a load in conditions where draining is hindered or partially limited, thus generating a high overall mobility. For liquefaction to take place, a soil must develop the conditions indicated previously, sufficient pressures in the pores to support the solid particles, this means that only soils with contractive tendencies can undergo liquefaction. The types of sediment most susceptible to liquefaction are deposits devoid of clay constituted mainly by sand and silt; however, in certain cases, gravel may also undergo liquefaction.

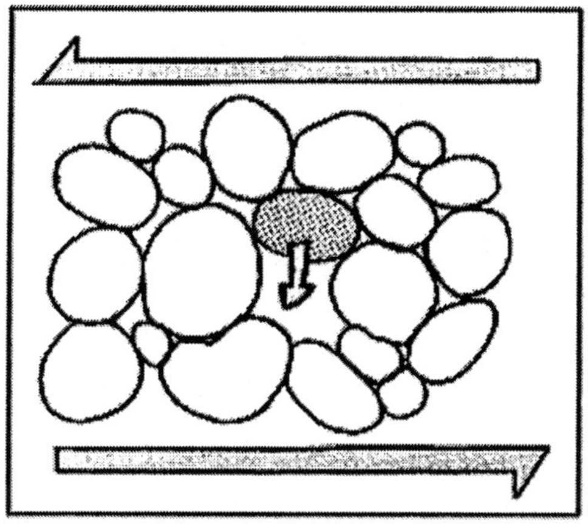

Figure 4.6: Diagram illustrating a layer of granules in a "loose" soil with a high degree of saturation subject to liquefaction. The strain due to shear stress (indicated by the horizontal arrows) and the stress caused by surface-sliding phenomena, distorts the granular structure, thus causing failures that reduce the interstitial spaces, as shown by the central arrow.

The process of liquefaction can also be explained as follows: the deformations that affect the debris mass in response to landslides and surface failures, propagate preferentially through the granular layers with high degrees of saturation, thus distorting their structure and causing instability and collapses of the particles (Figure 4.6). These failures in the granular layers of a soil with contractive behaviour cause the transfer of part of the applied load, previously supported in the friction contacts between grains, to the interstitial fluid, thus increasing pressure. This increase in pressure causes the sudden drainage of the interstitial fluid and therefore each increase in pressure can be continuously dissipated, as the liquid is able to flow away freely. However, if drainage is partially hindered, an abrupt and temporary increase in the interstitial pressure may take place. If interstitial pressure increases to the point in which it counterbalances the entire applied load, following liquefaction, the granular layer involved will temporarily behave as a plastic fluid rather than a solid. Therefore, in the case of liquefaction, part of the load, which was previously sustained by the stress due to mutual contact between grains, will now be sustained by the pressure in the interstitial fluid. When $\sigma_e \to 0$, one has $p \approx \sigma_n$, in other words, the interstitial pressure increases to the point of counterbalancing the entire applied load and liquefaction takes place (Figure 4.7).

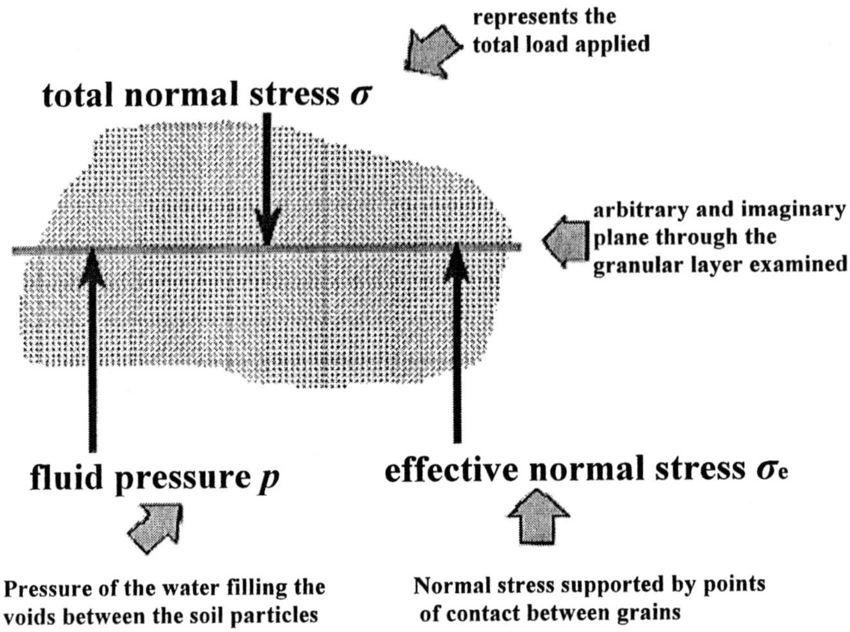

Figure 4.7: Diagram illustrating Terzaghi's principle of effective normal stress [47].

By differentiating the principle of effective stress, $\mathrm{d}\sigma_e = \mathrm{d}\sigma_n - \mathrm{d}p$, and assuming as an initial approximation that each variation in the total load is negligible, one has $\mathrm{d}\sigma_e = -\mathrm{d}p$. According to this equation, an increase in interstitial pressure has a corresponding decrease in the effective normal stress, also known as granular stress, the consequent loss of resistance is translated into a rapid and uncontrolled deformation (collapse) of the debris mass, thus favouring its passage to the state of fluid in movement.

4.3 Models on the mechanics of debris-flow triggering

The previous examination of the main types of triggering underlines that the factors that cause this phase of the process are the localisation of the trigger area and the main triggering mechanism (hydrodynamic actions, geotechnical causes and mechanical equilibrium). On a forecasting level, it is difficult to identify which of these mechanisms is the most probable or makes the largest contribution, in any case, any one of these types of triggering can include others, or rather, within the same event different trigger mechanisms may occur in different areas of the basin. In studies on the trigger conditions of cohesionless debris flows, following the appearance of a superficial water flow, analysis is concentrated on the study of the instability of the accumulation of materials that takes place following "saturation". A surface flow that flows along a slope following heavy rain, first "saturates" the debris layers, then mobilises them, thus causing the dispersion of the solid particles of the accumulation throughout the entire depth of the streamflow, which ultimately results in a debris-flow. In order to identify the mechanisms of involvement of the solid material within a liquid flow, we must analyse the role played by the various forces acting in the genesis of the motion. The approaches adopted for studying this matter are the traditional approach proposed by Shields [48], which analyses the instability of a mobile bed in streams characterised by bed-load or suspended load, through a transport mechanism that has been defined *individual particle motion* as it considers the equilibrium of a single particle, and that proposed by Takahashi [14, 21] which analyses the stability of an entire finite and "saturated" granular layer, subject to a surface flow of water, and for this reason it has been defined *massive motion.*

4.3.1 Shields' approach

From a phenomenological point of view, the bed motion of the solid particles due to the forces exerted by a streamflow can be perceived using the various conceptual schemes illustrated below. In any case, one must evaluate the stability of the particle considering the streamflow's driving force and resisting forces, caused by the immersed weight of the particle, and the consequent friction on the bed. When all other parameters are equal, with an increase in velocity, the hydrodynamic actions are able to overcome the stabilising actions, thus setting the particle in motion. The flow conditions in which the particle is on the boundary of movement, are called *conditions of critical* or *incipient movement.*

- Rubbing (friction)
  The particle rubs along the bed, the hydrodynamic forces must therefore overcome the resisting forces of sliding friction.

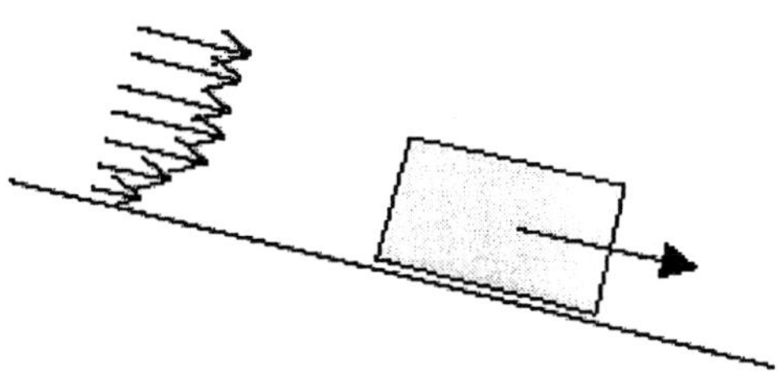

Figure 4.8: Bed motion of solid particles due to the forces exerted by the streamflow: rubbing scheme.

- Overturning (no friction)
  The particle is partially trapped on the bed, hydrodynamic forces must cause it to rotate on its point of rest.

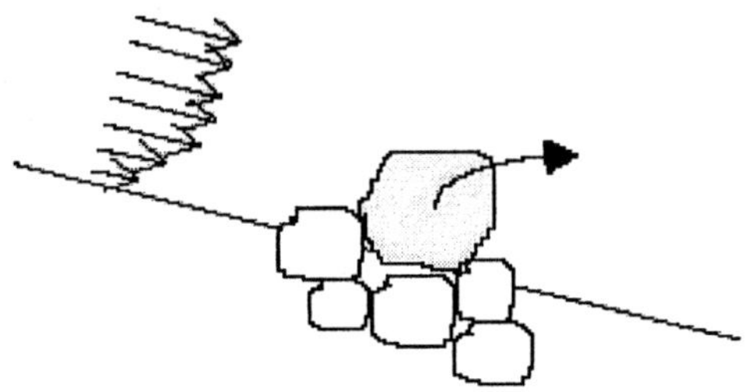

Figure 4.9: Bed motion of solid particles due to the forces exerted by the streamflow: overturning scheme.

- Jumps
  The particle is partially trapped on the bed and hydrodynamic forces must lift it up.

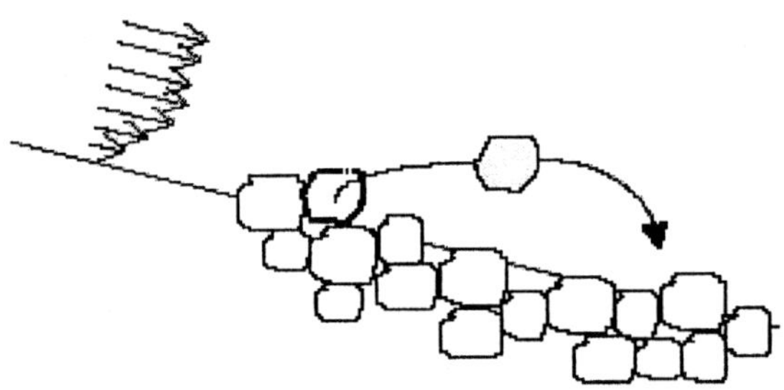

Figure 4.10: Bed motion of solid particles due to the forces exerted by the streamflow: overturning scheme.

If, for simplicity's sake, we refer to the sliding friction scheme alone, we can seek the conditions and therefore the relationships between the system parameters in conditions of incipient movement. The forces that act on the particle are as follows:

$$F_D = 0.5 C_D \rho_S \alpha_1 d^2 u^2 \tag{4.1}$$

$$F_L = 0.5 C_L \rho \alpha_1 d^2 u^2 \tag{4.2}$$

$$P - B = (\rho_s - \rho) g \alpha_2 d^3 , \tag{4.3}$$

with F_D the drag force, F_L the lift force, B Archimedes' force and P own weight (Figure 4.11).

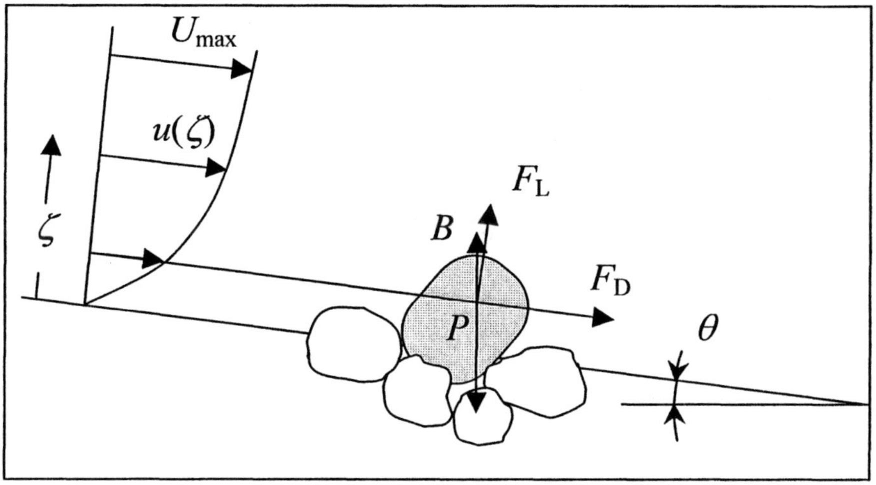

Figure 4.11: Diagram illustrating the set of forces acting around a solid element on a channel bed, in the sliding friction scheme.

In the previous expressions, u represented the local velocity of the streamflow around the particle, ρ is the density of the water flow and ρ_s is the density of the particle, C_D and C_L are the drag and lift coefficients, which depend on the shape of the grain (whose dimension is shown by d) and on the Reynolds number; the α_i coefficients are also linked to the shape of the grain and are considered constant and characteristic of the grain-size of the sediments.

In static equilibrium conditions, the normal R and tangential A friction force, must be, respectively, equal to:

$$R = (P - B)\cos\theta - F_L \tag{4.4}$$

$$A = (P - B)\sin\theta + F_D , \tag{4.5}$$

in which θ is the bed slope. The equilibrium yield condition (incipient movement) is therefore reached when:

$$A = R \tan\phi , \tag{4.6}$$

where ϕ is the grain's angle of internal friction. By expressing the terms of eqn (4.6) as follows:

$$(P-B)\sin\theta + F_{\mathrm{D}} = [(P-B)\cos\theta - F_{\mathrm{L}}]\tan\phi \\ \Rightarrow \\ F_{\mathrm{D}} + F_{\mathrm{L}}\tan\phi = (P-B)\cos\theta(\tan\phi - \tan\theta) \tag{4.7}$$

and by substituting the expressions for F_{D}, F_{L} and $(P-B)$ (eqns (4.1), (4.2) and (4.3)) into eqn (4.12), one obtains:

$$\frac{u^2}{gd\rho/(\rho_{\mathrm{S}}-\rho)} = 2\frac{\alpha_2}{\alpha_1}\cdot\frac{1}{(C_{\mathrm{D}}+C_{\mathrm{L}}\tan\phi)}\cdot(\cos\theta\tan\phi - \sin\theta). \tag{4.8}$$

Given the punctual velocity distribution of the streamflow $u(\zeta)$, which is based on the distance ζ from the bed (Figure 4.11), the shape of the profile imposes a relationship between the mean velocity U and the shear stress τ acting on the bed, the latter being linked to the shear rates near the bed. For a developed free surface flow, we can assume for the profile $u(\zeta)$ a "universal" type law such as $u(\zeta)/u^* = f(\zeta/d)$, to be expressed as a function of a constant calculated on the basis of the expression adopted for the velocity profile, which can be posed as $u/u^* = \alpha_3$. Equation (4.8) can be rewritten as follows:

$$\frac{u^{*2}}{gd(\rho_{\mathrm{S}}-\rho)/\rho} = 2\frac{\alpha_2}{\alpha_1\alpha_3^{\,2}}\cdot\frac{(\cos\theta\tan\phi - \sin\theta)}{(C_{\mathrm{D}}+C_{\mathrm{L}}\tan\phi)}. \tag{4.9}$$

Therefore, using the velocity profile form, it is possible to establish a relationship between the hydrodynamic conditions and the shear velocity u^*, the latter being linked to the shear stress on the bed $\tau_0 = \rho u^{*2}$, which are either known or can be easily calculated for a streamflow. Relationship (4.9) therefore becomes:

$$\frac{\tau_0/\rho}{gd(\rho_{\mathrm{S}}-\rho)/\rho} = 2\frac{\alpha_2}{\alpha_1\alpha_3^{\,2}}\cdot\frac{(\cos\theta\tan\phi - \sin\theta)}{(C_{\mathrm{D}}+C_{\mathrm{L}}\tan\phi)}. \tag{4.10}$$

The first term of eqns (4.9) and (4.10) is a dimensionless term that is known as *the Shields' number*:

$$\phi_{\mathrm{c}} = \frac{u^{*2}}{gd(\rho_{\mathrm{S}}-\rho)/\rho} = \frac{\tau_0/\rho}{gd(\rho_{\mathrm{S}}-\rho)/\rho}. \tag{4.11}$$

Equation (4.11) could be rewritten once more introducing also the grouping $\Delta = \rho/(\rho_{\mathrm{S}}-\rho)$, in any case the Shields' number is referred to as a dimensionless shear stress. The beginning of motion of bed material is a function of this dimensionless number. Given the characteristics of the streamflow (and therefore ρ and τ_0) and those of the sediments (Δ, ρ_{s}, d), in order to value particle stability we must compare ϕ with the critical threshold identified by eqn (4.11):

- for $\phi < \phi_c$ the grain is stable;
- for $\phi > \phi_c$ the grain is in movement.

It should be pointed out that the value of ϕ is linked to the dimension d of the granules: the transport phenomenon is selective and acts differently on grains of different dimensions. With an increase in the bed slope θ, one can observe, on the one hand an increase in streamflow velocity with a consequential increase in the drag and lift forces and the tangential component of the weight, which for gentler slopes has a negligible effect and, on the other, a decrease in the resisting forces and the normal component of the weight. Therefore, whereas on gentle slopes particle motion is essentially due to hydrodynamic forces, with steeper slopes (in Shields' scheme slopes of greater than approximately 20% are assumed), in the critical condition of the beginning of motion, the tangential weight component starts to have a non-negligible influence, in part due to the reduction in the normal component. This simple scheme of solid particle motion has been widely used by various authors in improving ordinary transport formulae and, despite its limits, it is extremely useful in understanding the forces at work in the area around the granules in critical initial motion conditions. The limits of this scheme are linked to its excessive simplification of the actions that act on the entire granular mattress of the channel bed, for this reason, in Shields' approach, the motion phenomenology is defined with the expression *individual particle transport* [14, 21]. However, it cannot be considered exhaustive in representing the mechanics responsible for triggering flows that can be classified as debris flows, in which it is essential to consider the equilibrium of the entire channel bed layer; an approach that can therefore be termed *massive motion* [14, 21].

4.3.2 Takahashi's approach

We will herein study the case of the motion of solid material on an infinitely extended channel bed, on which a constant flow runs out in uniform motion conditions. A simple analysis of stability is preferred in order to calculate the critical angle that causes the flow of a uniform layer h of a "saturated", cohesionless fluid–sediment mixture, with a density ρ_m; ϑ is the bed slope, ϕ the material's angle of internal friction and g is the gravitational acceleration.

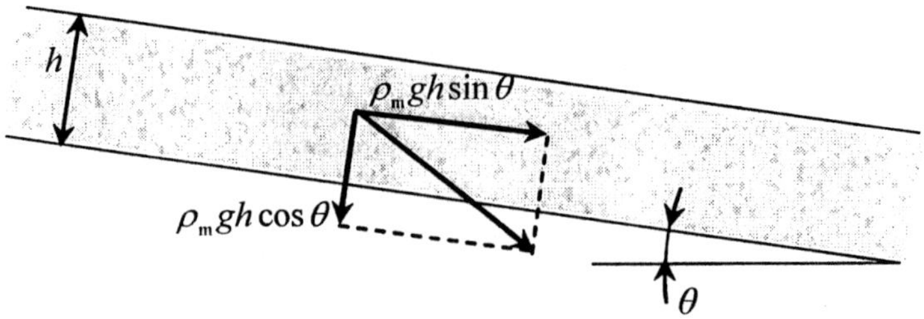

Figure 4.12: Definition diagram for the analysis of stability in Takahashi's approach.

According to the Mohr–Coulomb friction criterion, the threshold shear stress τ_L for a cohesionless material is given by:

$$\tau_L = \bar{\sigma}\tan\phi, \tag{4.12}$$

in which $\bar{\sigma} = \sigma - \sigma_p$ is the effective normal stress [47] (see Section 4.2.2), with:

$$\sigma_p = \rho_l hg\cos\theta \tag{4.13}$$

$$\sigma = \rho_m gh\cos\theta, \tag{4.14}$$

where ρ_l is the density of the intergranular fluid and ρ_m the density of the mixture, which is given by:

$$\rho_m = \rho_l + (\rho_S - \rho_l)\cdot C_v^*, \tag{4.15}$$

where ρ_s is the density of the solid particles and C_v^* is the maximum value of the volumetric solid concentration. The driving stress parallel to the surface τ, due to the weight of the mixture is:

$$\tau = \rho_m gh\sin\theta. \tag{4.16}$$

In the equilibrium condition it follows that:

$$\begin{aligned} \tau &= \tau_L \\ \rho_m gh\sin\theta &= (\sigma - \bar{\sigma})\tan\phi \\ \rho_m gh\sin\theta &= (\rho_m - \rho_l)gh\cos\theta\tan\phi \end{aligned}, \tag{4.17}$$

which gives:

$$\tan\theta = \left[\frac{C_v^*(\rho_S - \rho_l)}{C_v^*(\rho_S - \rho_l) + \rho_l}\right]\tan\phi. \tag{4.18}$$

As the coefficient of $\tan\phi$ is always less than 1, it therefore follows that $\tan\theta < \tan\phi$, therefore, the inclination required for the motion of a fluid–granule mixture is always less than that required for the sliding of a dry aggregate of similar granules. Bagnold [7] uses a similar approach referring, however, not to the angle of static friction ϕ, but to the dynamic one ϕ_d (Bagnold's limit angle criterion); this condition is, as we will see, an essential condition for triggering a debris flow.

Starting from this simple condition Takahashi [14], develops a study of cohesionless debris-flow triggering, which was subsequently reviewed, in order to take into account the influence of cohesion as well [20, 21]. The model assumes an infinitely vast uniform layer of cohesive material with a thickness of D, deposited on a channel bed with a slope θ; it is hypothesised that at the moment in which the water flow h_0 starts to run off, the porous spaces between the grains are completely "saturated" and there is a hydrostatic distribution of the pressures in the interstitial fluid. Another element considered is the depth co-ordinate a, whose origin is posed in correspondence with the granular surface which has a positive direction if turned towards the channel bed. Moreover, a_L represents the point in which resisting stress, linked to the internal friction contacts between grains, and the driving stress, linked to the applied load, assume the same value. Under these conditions, the stresses involved are as follows:

- active stress τ, represented by the component parallel to the surface of both the weight of the surface water and the weight of the "saturated" layer of sediments, with a thickness a:

$$\tau = (\rho_l h_0 + \rho_m a) g \sin\theta , \tag{4.19}$$

  and by substituting the expressions (4.15) for ρ_m, the density of the mixture, one obtains:

$$\tau = \left[\rho_l (h_0 + a) + a(\rho_s - \rho_l) C_v^{*}\right] g \sin\theta . \tag{4.20}$$

- The pressure in the pores, which is equal to the hydrostatic pressure σ_p at the height $(a + h_0)$:

$$\sigma_p = (a + h_0)\rho_l g \cos\theta . \tag{4.21}$$

- The component perpendicular to the surface, due to the weight:

$$\sigma = (\rho_m a + \rho_l h_0) g \cos\theta . \tag{4.22}$$

- Resisting stress τ_L, according to the Mohr–Coulomb friction criterion, also considering cohesion c:

$$\tau_L = c + \overline{\sigma} \tan\phi . \tag{4.23}$$

Thus referring, once again, to the concept of effective normal stress proposed by Terzaghi [47], we can write $\overline{\sigma} = \sigma - \sigma_p$, so, by substituting eqns (4.21) and (4.22), eqn (4.25) becomes:

$$\tau_L = c + C_v^*(\rho_s - \rho_l)ag\cos\theta\tan\phi\,. \tag{4.24}$$

Assuming a linear distribution, the profile of the shear stress within the granular material should fall into one of the two boundary cases:

1. $\boxed{\dfrac{d\tau}{da} \geq \dfrac{d\tau_L}{da}}$ Thus by deriving eqns (4.20) and (4.24) and substituting them into this condition, one obtains (Figure 4.13):

$$\tan\theta \geq \frac{C_V^*(\rho_S - \rho_l)}{C_V^*(\rho_S - \rho_l) + \rho_l}\tan\phi \tag{4.25}$$

It should be noted that in almost static conditions, this relationship is nothing more than the limit angle expressed in eqn (4.18).

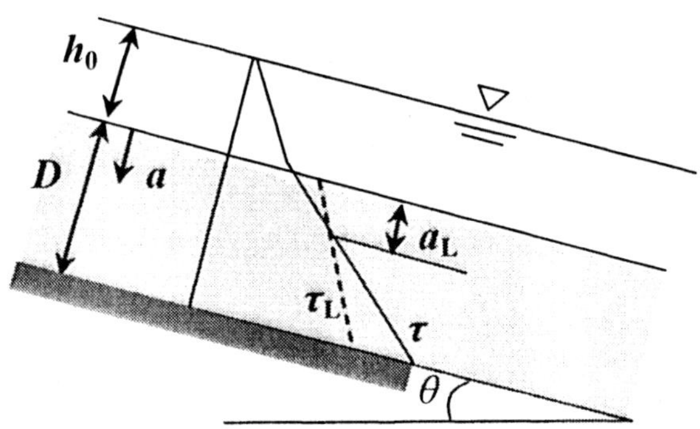

Figure 4.13: Diagram describing the trend of the applied stress τ and the internal resistance stress τ_L, in a granular layer subject to a water flow h_0: case $\dfrac{d\tau}{da} \geq \dfrac{d\tau_L}{da}$.

This condition is satisfactory from three configurations, in relation to the position that the point of intersection between the straight line τ_L and the straight line τ assume as a function of coordinate a:

- $a_L \leq 0$, which means that $\tau \geq \tau_L$, corresponding to a debris flow without rigid plug;

- $0 < a_L < D$ corresponding to a debris flow with rigid plug (case shown in Figure 4.13);
- $a_L \geq D$, which means that $\tau \leq \tau_L$, a case in which a debris flow cannot take place, as the bed is stable.

In Takahashi's first approach, cohesion, corresponding to the intercept of straight line τ_L, was not considered (c=0), a somewhat frequent condition in a mass of sediments that has accumulated over a long period of time [21]. In this way, subcase $\tau \geq \tau_L$ only can occur; by substituting into eqn (4.27) eqns (4.20) and (4.24), for $a = a_L$, we have:

$$\tan\theta \geq \frac{C_v^*(\rho_s - \rho_l)}{\rho_l\left(\frac{h_0}{a_L}+1\right)+(\rho_s-\rho_l)C_v^*}\tan\phi . \qquad (4.26)$$

Above $a = a_L$ because $\tau \geq \tau_L$ only in the interval of values of a between a_L and D, and, for a datum h_0, $\tan\theta$ is minimal because $a = a_L$. In comparison with eqn (4.27), this relationship has the additional term h_0/a_L which describes the role that the appearance of a surface flow has in distinguishing landslides from debris flows. In fact, if this flow is not able to reach the surface of the mixture, one has h_0=0, the two conditions are equivalent and a landslide takes place because the structure of the bed is maintained, given that in the surface layers, resisting stress prevails over active stress. In other words, it can be assumed that a debris-flow only occurs when a flow of surface water appears on top of the accumulation of sediments [14].

2. $\boxed{\frac{d\tau}{da} < \frac{d\tau_L}{da}}$ In the same way one obtains (Figure 4.14):

$$\tan\theta < \frac{C_v^*(\rho_s - \rho_l)}{C_v^*(\rho_s - \rho_l)+\rho_l}\tan\phi . \qquad (4.27)$$

Despite the fact that this condition does not encounter the criterion of the limit angle expressed by eqn (4.18), a debris flow may, however, take place in the upper part of the accumulation ($a \leq a_L$), which is unstable due to $\tau \geq \tau_L$. If a_L is lower than the representative diameter of the sediments (d), the layer is stable, despite the previous condition and only the superficial particles are transported by the hydrodynamic force of the streamflow. This cannot be the case of a debris-flow, but rather corresponds to ordinary bed-load or suspended load.

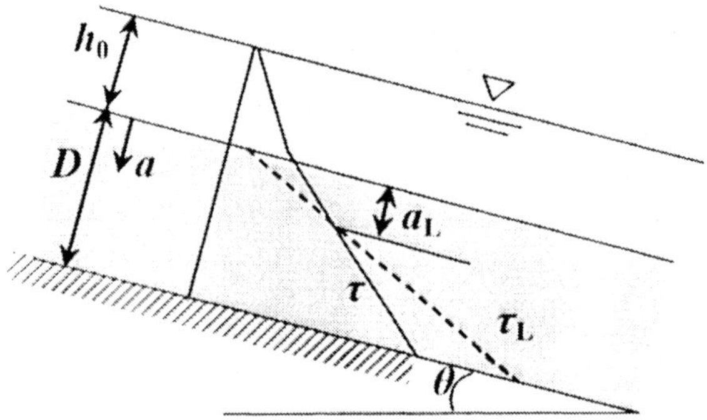

Figure 4.14: Diagram describing the trend of applied stress τ and internal resisting stress τ_L, in a granular layer subject to a water flow h_0: case $\dfrac{d\tau}{da} < \dfrac{d\tau_L}{da}$.

In order to determine a debris flow, it is necessary that $a_L \geq d$ (for example, $a_L = nd$ with $n \geq 1$); thus substituting d instead of a_L in eqn (4.26), with the condition $a_L \geq d$, one obtains:

$$\tan\theta \geq \frac{C_v^*(\rho_s - \rho_l)}{C_v^*(\rho_s - \rho_l) + \rho_l\left(\dfrac{h_0}{d} + 1\right)} \tan\phi . \tag{4.28}$$

Supposing that this condition also takes place, if a_L is much lower than the height of surface flow h_0, the particles are not able to disperse throughout the entire depth of the flow, but concentrate in the lower part, thus causing an *immature debris flow*; and it is therefore necessary that $a_L \geq kh_0$, with k a numerical constant whose experimental value is equal to 0.7 [14]; by substituting a_L with kh_0 in eqn (4.26), with the condition $a_L \geq kh_0$, one obtains:

$$\tan\theta \geq \frac{C_v^*(\rho_s - \rho_l)}{C_v^*(\rho_s - \rho_l) + \rho_l\left(\dfrac{1}{k} + 1\right)} \tan\phi . \tag{4.29}$$

In conclusion, by requiring that both eqns (4.28) and (4.29) are satisfied simultaneously (which is an even more restrictive condition) together with eqn (4.27), one obtains the criterion corresponding to the second case for debris-flow triggering:

$$\frac{C_v^*(\rho_s - \rho_l)}{C_v^*(\rho_s - \rho_l) + \rho_l\left(\dfrac{1}{k} + 1\right)} \tan\phi \leq \tan\theta < \frac{C_v^*(\rho_s - \rho_l)}{C_v^*(\rho_s - \rho_l) + \rho_l} \tan\phi . \tag{4.30}$$

The analysis conducted thus far is valid for cohesionless mixtures; if we were to consider the cohesion of a granular bed, it would be necessary to introduce an additional term to the numerator of criteria (4.26) and (4.30), and according to the relative weight of this term and ϕ, the inequalities could be altered as they are deduced using the simple condition $\tau \geq \tau_L$. One solution to this problem would be to refer to the criterion (4.25), obtained using the expression of the stresses and is valid as the lower limit for the trigger condition for a cohesive, almost static flow and for a cohesionless one.

Using the data of Takahashi [14], C_v^*=0.7, ρ_s =2.6 g/cm^3, ρ_l =1.0 g/cm^3, k=0.7 and tanϕ=0.8, one obtains critical triggering slopes of between 14.5° and 22.9°. By drawing a graph of the inequalities obtained, it is possible to identify the regions in which the various sediment-transport processes exist (Figure 4.15). The domain indicated by A corresponds to the absence of particle motion, and is delimited by a curve that represents the triggering condition of ordinary bed–load transport. The domain indicated by B corresponds to the movement of the single particles (Shields' approach) under the action of hydrodynamic forces in ordinary bed-load; the area indicated by C corresponds to the sediment movements, in response to gravity and hydrodynamic forces, which represents the immature debris flow in which, with a dense granular flow, a pure liquid current flows. The area is delimited by two curves corresponding to eqn (4.28) for increasing values of the ratio h_0/d; one can observe that the hydrodynamic effect tends to diminish as the thickness of the layer in movement increases. D is the region corresponding to debris-flows, in which the granules are dispersed throughout the entire thickness, E is the transition domain between debris-flows and landslides and F corresponds to a situation of purely static instability for sediment beds, irrespective of the presence of a liquid streamflow.

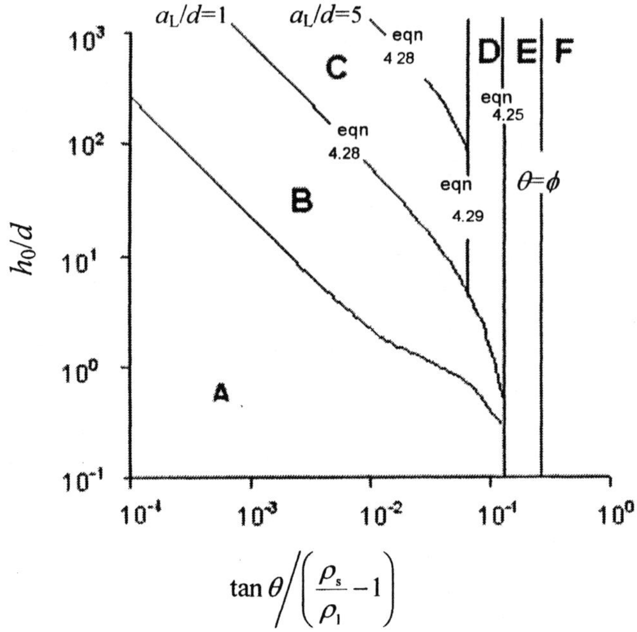

Figure 4.15: Trigger criteria for the occurrence of the various types of sediment transport; the curves are obtained with the following data: C_v^*=0.7, ρ_s=2.6 g/cm^3, ρ_l=1.0 g/cm^3, k=0.7 and $\tan\phi$=0.8. [14]. The domain indicated with A corresponds to the absence of motion, that indicated with B corresponds to ordinary bed-load, the domain indicated with C corresponds to immature debris flows, domain D to debris-flows, domain E corresponds to the transition between debris-flow and landslide and lastly, domain F corresponds to a situation of purely static instability.

4.3.3 Recent developments

Studies into the triggering conditions of a cohesionless debris flow following the appearance of a surface water flow have recently introduced the effect of the positioning of a single particle in relation to the average surface through its degree of exposure e [5].

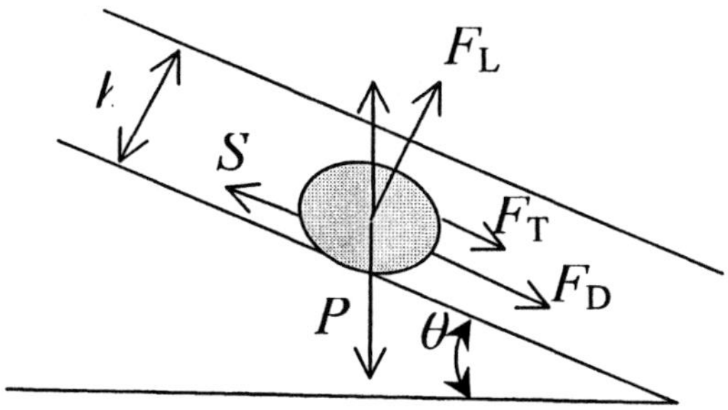

Figure 4.16: Forces acting on a single particle belonging to a bed with an inclination of θ on which a water flow with a thickness of h runs off [5]. B Archimedes' force, P own weight, F_L lift, F_D drag, F_T and S forces depending on the particle's degree of exposure.

In this section we will demonstrate how both of the previously illustrated approaches are inadequate with experimental data, as they do not consider the nature of the bed and as the relationships obtained are not directly comparable. Returning to eqn (4.11), which represents the critical condition for triggering motion in the Shields' approach, we can look for a similar expression in Takahashi's model. For $a=a_L$, by equalling eqn (4.19), which represents the expression of active stress, and eqn (4.24), which represents that of resisting stress, in which, however, cohesion is neglected (c=0), and assuming for convenience's sake that $\rho_l=\rho$, density of the water flow, $\tau_0 = \rho g h_0 \sin\theta$ stress at the channel bed (in the hypothesis of steady and uniform flow), one obtains:

$$\frac{\tau_0/\rho}{g a_L(\rho_s - \rho_l)/\rho} = C_v^* \cos\theta \tan\phi - \left(C_v^* + \frac{\rho}{\rho_s - \rho_l}\right)\sin\theta\,, \tag{4.31}$$

in which ρ_s is the density of the granules, g is gravitational acceleration, C_v^* is the maximum volumetric concentration of the sediments, θ is bed slope, ϕ is the granules' angle of internal friction, a_L is the coordinate perpendicular to the channel bed and measured from the surface of the granular layer, of the point in which applied stress and resistive stress are equal. If we then introduce the shear velocity $u^{*2} = \tau_0/\rho$, and remembering that Takahashi poses a_L=nd with n≥1 and d granule diameter, we can write a similar expression to eqn (4.11):

$$\frac{u^{*2}}{gd(\rho_s - \rho)/\rho} = n\left[C_v^* \cos\theta \tan\phi - \left(C_v^* + \frac{\rho}{\rho_s - \rho}\right)\sin\theta\right]. \tag{4.32}$$

By neglecting multiplicative constants, one observes that the two relationships only have common solutions if $C_v^* = 1$ and $C_v^* = -\rho/(\rho_s - \rho)$ at the same time.

In order to overcome this lacking supposition, from the point of view of a feasible extension of the models, Armanini and Gregoretti [5] underline the inadequacy of the approaches could be directly due to having neglected the nature of the bed. By introducing parameter *e*, the degree of exposure of a single particle to the streamflow that varies from 0 to 1 when the particle is completely hidden and completely exposed, respectively. Consequentially, the drag and lift forces will be altered according to two suitable functions of the degree of exposure, $f_1(e)$ and $f_2(e)$; it will also be necessary to introduce two new contributions in the balance of the forces: F_T, the resistive force due to action of the bed on the projection of the exposed part of the particle, which is also a function of the degree of exposure through $f_3(e)$, and S, the force acting on the lower, nonexposed part of it depending on $f_2(e)$.

$$F_D = 0.5C_D \rho_s \alpha_1 d^2 u^2 f_1(e) \tag{4.33}$$

$$F_L = 0.5C_L \rho \alpha_1 d^2 u^2 f_1(e) \tag{4.34}$$

$$S = \rho g(1-\alpha_2) d^3 \sin\theta [1 - f_2(e)] \tag{4.35}$$

$$F_T = \alpha_1 d^2 \tau_0 f_3(e), \tag{4.36}$$

in which u is the local velocity of the streamflow around the particle, C_D and C_L are the drag and lift coefficients and the α_i are coefficients linked to the shape of the grain.

In the case of a spherical particle, eqn (4.6) is altered as follows:

$$F_D + (P-B)\sin\theta + S + F_T = (P-B)\cos\theta\tan\phi - F_L \tan\phi . \tag{4.37}$$

By hypothesising suitable expressions for these functions and remembering that $\Delta = \rho/(\rho_s - \rho)$, it is possible to obtain a similar equation to eqns (4.11) and (4.32):

$$\frac{u^{*2}}{gd\Delta} = 2\frac{\alpha_2}{\alpha_1 f_1 \alpha_3^2} \cdot \frac{1}{(C_D + C_L \tan\phi) + \alpha_1 f_3} \cdot \left[\cos\theta\tan\phi - \sin\theta - (1-f_2)(1-\alpha_2)\cdot\frac{\sin\theta}{\Delta\alpha_2}\right], \tag{4.38}$$

which, thanks to the introduction of the degree of exposure *e*, represents a more general extension of the conventional theories on incipient motion conditions [5].

4.4 Conclusions

A critical analysis has been made of the various models currently available for describing the mechanisms of debris-flow triggering. Various schemes of involvement and transport of solid material within a flow running off a natural channel bed have been considered. The study of the types of stress acting on an initially immobile granular mass, irrespective of the type of moving mechanics considered, individual particle movement or massive motion, underlines the importance of the bed slope, which determines the net variation in the driving and resisting forces, in the passage between ordinary solid-transport phenomena and debris-flows (see Figure 4.15). On gentle slopes (less than 5%) one can observe the movements typical of the first mechanism, whereas as the slope increases (approximately 20%) one observes a transition towards phenomena typical of debris-flows, in which the movement of the entire layer of the channel bed is clearly predominant. This illustrates why the phenomenology indicated as a debris-flow is often described as a flow subject to gravity, given that, at the corresponding inclinations, the force of gravity assumes a decisive role. It must, however, be noted that different types of transport coexist within the flow depth as sediment characteristics and flow behaviour vary. This obviously means that under the simplified conditions mentioned above, which consider the case of an infinitely vast bed in steady conditions, excluding other trigger mechanisms, as the aim was to examine in particular the most significant aspects in debris-flow phenomenon genesis, in order to understand the differing mechanics than liquid flows with ordinary solid transport.

Chapter 5

The state of the art of modelling debris-flow dynamics of motion

5.1 Introduction

This chapter provides a critical analysis of the various models proposed in the literature on the mechanics of a steady and uniform debris flow, in an attempt to identify those aspects that are as yet not fully understood and that inevitably condition the results that can be obtained from the application of mathematical models, thus giving weight to the opinion that only a more thorough knowledge of the parameters involved can lead to more realistic simulations of debris-flow events in the future.

The development of a debris flow along a channel has three separate and heterogeneous regions: the gravely head, a viscous body, and a turbulent tail. These characteristics cause a flow situation that is far removed from any possible ideal fluid model. It would therefore seem apt to describe its behaviour with three distinct constitutive relationships, however, from a practical point of view, in the absence of a lack of systematic methods for determining the properties of the mixture in each region; this further attempt may not give the expected results [1]. Therefore, in order to fully understand the mechanics of a debris-flow, it would seem suitable, at least initially, to adopt a simplified treatment. There are four key characteristics in momentum transport within the phenomenon: the viscosity of the interstitial fluid, turbulence, sliding friction, and the collision of the particles. The dominance of these properties in debris-flows has led to the qualitative classification in six different regimes [49]. The classification of the typical debris-flow regimes proposed by Jan and Shen [49] systematically deals with the various types of debris-flows that can take place, as it highlights how the prevalence of a certain set of stresses is determinant in characterising the phenomenology of this study. This shows, as does any traditional rheological model (or models created by combining the constitutive equations of base models) can be considered the most suitable for a certain type of streamflow. Therefore, from a historical point of view, we can justify the development of apparently conflicting models for different geographical areas for debris-flow studies; for example, whereas Chinese studies generally follow the viscoplastic hypothesis, Japanese ones mainly follow a mainly dilatant line [1]. We will therefore analyse the foundations of debris-flow mechanics, reviewing the main models available in the literature on the various regimes; starting from the constitutive equations we will extrapolate the velocity characteristics for a steady and bidimensional debris-flow, treating it as a uniform mixture. We will then move on to examine the layered models that have recently been introduced, as well as the hydraulic approach to the dynamic problem.

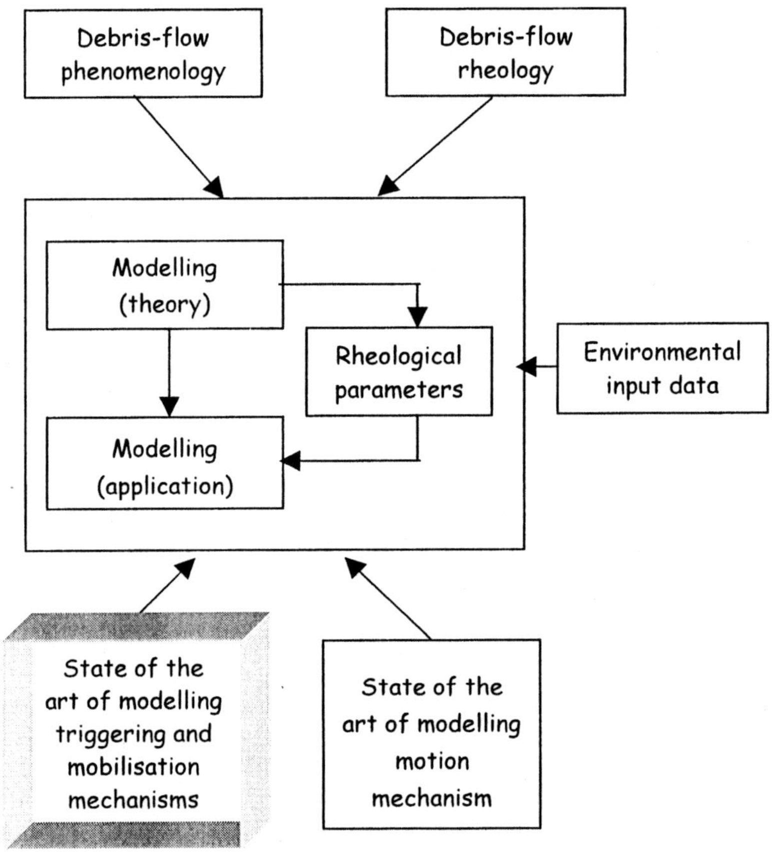

Figure 5.1: In the chart, showing a breakdown of the study approach adopted in this book, the grey frame highlights the analysis field of the state of the art of modelling motion mechanisms, the topic that will be dealt with in this chapter.

5.2 Dynamics of debris flows

Research work on debris-flow modelling has traditionally specialised in viscoplastic flows on the one hand (Johnson, [9]) and granular flows on the other (Bagnold, [7]; Savage, [17]). Each type of model postulates a single rheological equation between stress and strain; however, this assumption conflicts with both recorded data, which shows that solid and fluid stress vary asynchronously in a debris-flow, and with the developments in interstitial pressure and granular temperature [3]. Because a debris-flow essentially constitutes a multiphase system, any attempt at modelling this phenomenon that assumes as a simplified hypothesis homogeneous mass and constant density, conceals the interactions between the phases and excludes the possibility of investigating the problem's intrinsic phenomena, such as phase diffusion. It is therefore preferable that models that introduce a separate description for each phase, but that also consider their integration, will be adopted. On a microscopic level (see Section 3.5.1,), by means of inelastic collisions, granules that are nonuniformly distributed lose energy following each contact; therefore, by travelling at the mean velocity of the flow and exchanging momentum with the fluid they also acquire chaotic components of fluctuation. The consequential moving together of the grains can, as a preliminary approximation, cause a local increase in intergranular pressure and a runoff of the interstitial fluid in the adjacent pores [42], with consequent energy dissipations. At the same time the fluid dampening the collisions should cause further energy loss in momentum transfer. This microstructural dynamic, centred on multiparticle collisions and combined with statistical approximations (Pasquarell, [50]), has been one of the main approaches used in modelling dry granular mixtures, with the aim of determining the continuous properties of the medium by a statistical mean, starting from the analysis of the individual collision. One of the problems in the direct application of these results to debris-flows is that they have been calculated for dry mixtures, without considering the presence of the interstitial fluid and its interaction with the granular fraction. Despite the fact that it represents a milestone in debris-flow research, Bagnold's theory [7], suffers the same limitation as it assumes purely geometric and kinematic conditions. Other approaches (Goodman and Cowin, [51]; McTigue, [52]; Savage and Sayed, [53]) have been developed by the mechanics of continuous media. The modelled medium is therefore composed of a mixture of granules and fluid that behaves as a single entity; however, by neglecting the microstructure of the continuum, it is necessary that the parameters introduced in these theories are extrapolated from experiments. Generally speaking, it can be said that a good constitutive model for debris-flows should incorporate its viscous nature, fluid–sediment interactions and friction.

As mentioned earlier, the development of a debris flow along a channel has three separate regions of inhomogeneity: the gravely head, a viscous body, and a turbulent tail. It would therefore seem apt to describe its behaviour with three distinct constitutive relationships, however, from a practical point of view, in the absence of a lack of systematic methods for determining the properties of the mixture in each region; this further attempt may not give the expected results. There are four key characteristics in momentum transport within the phenomenon: the viscosity of the interstitial fluid, turbulence, sliding friction and the collision of the particles. The dominance of these properties in debris-flows has led to the qualitative classification in six different regimes [49]. The table below shows the various flow regimes, highlighting the peculiarities that the streamflow assumes in a given regime and justifying for each one the adoption of attributes such as granular, muddy, and stony flow alongside the general debris-flow terminology.

Table 5.1: Qualitative classification of debris flows in six different regimes by Jan and Shen [49].

FLOW REGIME	TYPE OF DEBRIS-FLOW	MAIN EFFECTS
1 FRICTION REGIME 2 COLLISIONAL REGIME 3 FRICTION–COLLISIONAL REGIME	*Granular flow*	➤ Negligible interstitial fluid ➤ Dominant granular interactions
4 MACROVISCOUS REGIME 5 VISCOPLASTIC REGIME	*Muddy flow*	➤ Dominant interstitial fluid ➤ Negligible granular interaction
6 COLLISIONAL VISCO PLASTIC REGIME	*Stony flow*	➤ Dominant granular and interstitial fluid interactions

5.3 General nature of the problem

Despite the fact that they contain solid particles, debris flows are usually schematically treated as the motion of a continuum. This approximation makes the equations for mass conservation and momentum for debris-flows similar to those for the motion of a generic fluid.

$$\frac{d\rho_m}{dt} + \rho_m \frac{\partial u_i}{\partial x_i} = 0 \tag{5.1}$$

$$\rho_m \frac{\partial u_j}{\partial t} + \rho_m u_k \frac{\partial u_j}{\partial x_k} = \rho_m f_j + \frac{\partial \sigma_{i,j}}{\partial x_j}, \tag{5.2}$$

where:

ρ_m is the density of the mixture with $\rho_m = (1-C)\rho_f + C\rho_s$, in which ρ_f and ρ_s are, respectively, the density of the interstitial fluid and that of the solid particles and C is the volumetric solid concentration;

t is time;

x_k is the generic Cartesian coordinate, with k=1, 2, 3 that represent directions x, y, z, respectively;

u_k is the generic velocity component;

f_j is the generic component of body force;

$\sigma_{i,j}$ is the stress tensor, in which the first subscript indicates that the stress component acts on the plane x_i=*cos* and the second subscript indicates that it is directed along direction x_j. The stress tensor $\sigma_{i,j}$ is generally expressed as $\sigma_{i,j} = -p\delta_{i,j} + \tau_{i,j}$; where p is pressure, $\tau_{i,j}$ is the tangential tensor and $\delta_{i,j}$ *Kronecker*'s delta function.

In the hypothesis of a uniform, steady and bidimensional debris flow along a plane with an inclination of θ, the motion equation, eqn (5.2), projected along x, suitably simplified and integrated between a generic height y and the free surface of the streamflow (y=h), gives $\tau_{xy}(y) = \rho_m g(h-y)\sin\theta$, in which g is the gravitational acceleration.

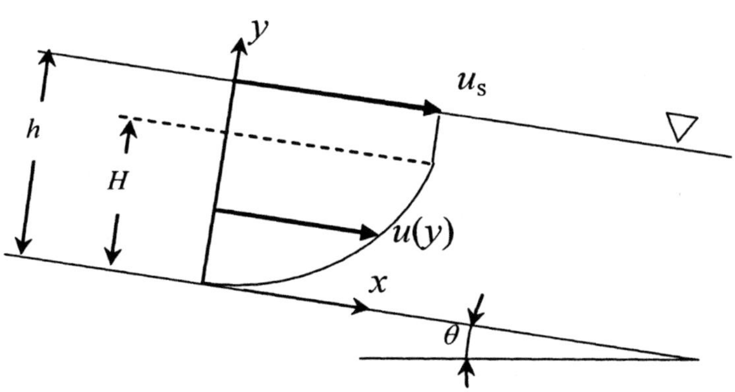

Figure 5.2: Definition diagram for a uniform, steady and bidimensional debris flow [49].

Below, the main constitutive equations available in the literature for each flow regime will be presented, which provide a suitable expression for $\tau_{xy}(y)$ to be introduced in the previous equation and therefore the characteristics of the velocity $u(y)$ will be obtained; Figure 5.2 shows a diagram defining the problem.

5.4 Models on the dynamics of debris flows in uniform motion

5.4.1 Frictional regime

In general, momentum transport during a very slow motion of granular material essentially originates from the mutual contact between the particles, whereas the action of the interstitial fluid and the effect of the dynamic stresses are negligible [49]. For approximately uniform, natural sand, this regime can take place for volumetric solid concentration values between 0.56 and 0.51 [33]. In this kind of system, when the particles move, they remain stuck closely together and deformation is low; the particles maintain contact for long periods, while they slide relative to one another and the transfer of momentum is mainly caused by the persistent rubbing of particles against adjacent ones. The stresses are therefore not governed by the applied strain, but mainly by peripheral stresses. In this case, the flow regime is called the frictional regime, plastic regime or quasistatic regime [9, 17], and debris-flow mixtures in this category show the typical properties of a plastic material. In recent decades, attempts have been made to apply certain theories on the plasticity of metals for the development of a valid theory for granular materials; an essential ingredient of these theories is the close dependency of all parameters on the pressure in the limit condition, as in the theory of Mohr–Coulomb, which states that at a point of a granular solid, plastic failure takes place, when on each section, shear stress τ and normal stress σ satisfy the relationship:

$$\tau = \tau_c + \sigma \tan\phi , \qquad (5.3)$$

in which τ_c and ϕ are cohesion and the angle of internal friction, respectively, which describes the friction resistance of the mass of granular materials. The normal stress component σ is, in general, dependent on the shear rate. Usually ϕ is considered constant for a granular accumulation of this type, although it is well known that it could depend on deformation and solid concentration. Certain studies (Hungr and Morgenstern, [54]), that examine the problem of mechanical fluidification in the frictional regime, show that the angle of internal friction of the granular material does not have a systematic dependency on normal stress and strain. However, the yield condition (eqn (5.3)) does not directly give any information on the kinematics of the motion of granular materials, because it is only valid in quasistatic regimes; it can only be extrapolated from velocity equations. The frictional model is used with a certain degree of success in forecasting the distributions of stresses and velocities of a granular material in laboratory experiments involving quasistatic regimes [53]. However, the applicability of the frictional model to a granular flow with a high deformation gradient is uncertain [55], given that the same mechanical fluidification (see Section 3.4.3) brings in these conditions an overturning in the Coulombic law of friction due to a change in the type of contact that exists between the particles, which causes high mobility of granular mixtures even on gentle slopes.

5.4.2 Collisional regime

In contrast to the frictional regime, when solid particles are sufficiently well spaced and deformation is rapid, contacts are brief and more similar to strikes and the momentum is mainly transferred by particle collision [49]. In this case, the particles tend to be dispersed throughout the entire depth of the streamflow and the solid fraction is represented by lower concentrations. In this case, the flow regime is termed the collisional regime, completely dynamic regime or the regime of grain inertia [7, 17, 56]. It has been found that the momentum transferred by particle collision is proportional to the square of the shear rate [7, 46, 57], according to the following general constitutive equation:

$$\tau = \alpha\left(\frac{\mathrm{d}u}{\mathrm{d}y}\right)^2, \tag{5.4}$$

where α is a coefficient that closely depends on density, dimension, grain-size, and the concentration of solid particles, as well as the angle of internal friction of the granular mixture. Equation (5.4) was initially proposed by Bagnold (1954) [7] and is usually known as the dilatant or dispersive model (see Section 5.4.2.1). Following experimental work by Bagnold, many researchers, such as Savage and McKeown [58], Savage and Sayed [53], Hanes and Inman [59], have conducted similar experiments; their results correlate with those obtained by Bagnold qualitatively but not quantitatively, and there is an order of magnitude difference between theoretical and experimental results [46]. If α is constant, the velocity components for a steady, uniform and bidimensional flow along an inclined plane in a collisional regime, are obtained using eqn (5.4) and are shown in Table 5.2; in agreement with Figure 5.2, θ is the angle of the inclined plane, ρ_m the density of the mixture, g gravitational acceleration, h is the depth of the flow, u_s is the velocity of the free surface, U is the mean velocity calculated by integrating on the flow depth and then dividing by the depth. One can observe that the mean velocity is equal to three fifths of the free surface.

Table 5.2: Velocity components for a uniform, bidimensional debris flow in the collisional regime.

Velocity distribution	$u = \frac{2}{3}\sqrt{\frac{\rho_{m} g \sin\theta}{\alpha}}\left[h^{1.5} - (h-y)^{1.5}\right]$
Surface velocity	$u_{s} = \frac{2}{3}\sqrt{\frac{\rho_{m} g \sin\theta}{\alpha}}h^{1.5}$
Velocity defect	$\frac{u_{s} - u}{u_{s}} = \left(1 - \frac{y}{h}\right)^{1.5}$
Average velocity	$U = \frac{2}{5}\sqrt{\frac{\rho_{m} g \sin\theta}{\alpha}}h^{1.5} = \frac{3}{5}u_{s}$

5.4.2.1 Bagnold's dispersive model

Despite their fundamental importance, the experiments conducted by Bagnold (1954) [7], stimulated by his interest in the phenomena linked to solid bed–load transport in water flows, went unnoticed for some time. These tests were carried out for various volumetric solid concentrations on uniform spherical particles with diameters equal to 1.32 mm, suspended in a Newtonian fluid with equal solid and liquid density, within a system of coaxial cylinders, of which the outermost was rotating. Bagnold distinguished two types of boundary behaviour: macroviscous and granuloinertial behaviour, which were shown in three flow regions, according to the value assumed by a dimensionless group, which represents the ratio between the inertial and viscous forces, subsequently defined as the Bagnold's number:

$$N_{\mathrm{BAG}} = \frac{\lambda^{1/2}\rho_{f} d^{2}}{\mu}\frac{\partial u}{\partial y}, \tag{5.5}$$

in which μ and ρ_f are the viscosity and density of the interstitial fluid, λ is the linear concentration of the particles defined as the ratio between particle diameter and the mean free distance between them, d is the diameter of the particles, y is the Cartesian coordinate in the direction normal to the flow, positive upwards, and u is the velocity component in the y direction.

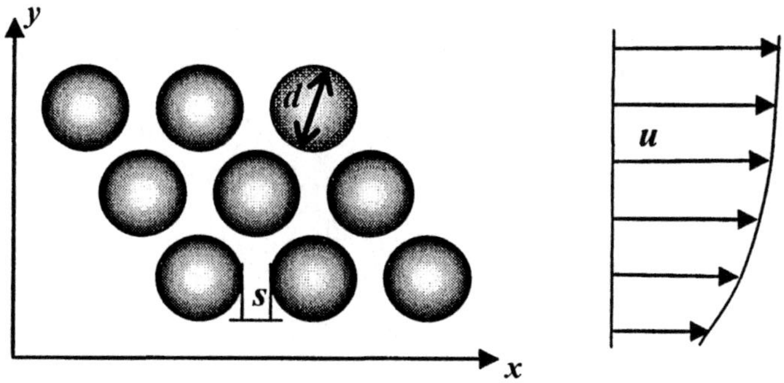

Figure 5.3: Definition diagram in Bagnold's analysis [7] for a granular mixture subject to deformation.

If the mean distance between the centres is bd and the free distance is s (Figure 5.3), it follows that $b = 1 + s/d = 1 + 1/\lambda$, having posed $\lambda = d/s$. The volumetric solid concentration can therefore be expressed in terms of λ as $C = \frac{C^*}{b^3} = \frac{C^*}{(1+1/\lambda)^3}$, where C^* is the maximum volumetric solid concentration that one obtains for $\lambda=\infty$ (for uniform spheres $C^*= 0.7405$). For $N_{BAG} <40$ (that is for low shear rate values) viscosity is dominant and normal and shear stress are linearly dependent on the shear rate: Bagnold called this flow regime the macroviscous region. For $N_{BAG} >450$ (corresponding to high shear-rate values), the interstitial fluid plays a minor role and the main effects are linked to interparticle interactions, and therefore the stress depends on the square of the shear rate: Bagnold called this flow regime the inertial region; between these two regions is the transition region in which the viscous and inertial effects are both important. The most interesting phenomenon to emerge from these experiments in each region considered is linked to the presence of a normal stress in a radial direction, termed by Bagnold as dispersive pressure, because it causes dilation within the material, proportional to the shear stress. Bagnold explained this characteristic, which was similar to the quasistatic behaviour of a cohesionless Coulombic material from a statistical point of view, as the result of a preferential anisotropy existing in the spatial distribution of the particles. The concept of dispersive force can be generally illustrated with a simple example [4]: as shown in Figure 5.4, a particle P situated in point 1 in the moment t_1 moves at a velocity V above P_1, reaching point 2 after a collision with P_1, which causes an abrupt variation in its velocity with regards to both velocity and direction, thus becoming V'; this obviously causes an acceleration. Mean acceleration in the interval (t_2-t_1) equals:

$$\vec{a} = \frac{\left(\vec{V}' - \vec{V}\right)}{t_2 - t_1} = \frac{\Delta\vec{V}}{\Delta t}. \tag{5.6}$$

In agreement with the second law of dynamics, particle P must be subject to the action of a force, whose mean value can be evaluated with:

$$\vec{f} = M \cdot \vec{a} = M \cdot \frac{\left(\Delta\vec{V} \cdot \vec{i}\right)\vec{i} + \left(\Delta\vec{V} \cdot \vec{j}\right)\vec{j}}{\Delta t}, \tag{5.7}$$

in which the unit vectors i and j in the horizontal and vertical direction respectively and the mass M of the particle appear. The vectorial components of ΔV are also represented in Figure 5.4; by multiplying them in scale by M it is possible to obtain the horizontal and vertical components of the force f and the dispersive force is quite simply the vectorial sum of these components. The dispersive force component in the flow direction and in the unit area, indicated by T, is defined as the dispersive shear stress, whilst the component of the dispersive force in a vertical direction and in the unit area, indicated by P, is called dispersive pressure. Dispersive pressure differs from the pressure in the fluid, in fact, the action of the pressure of the interstitial fluid on the cohesionless granules of the channel bed can effectively be transmitted to the particles thanks to the presence of cavities between particles "saturated" with fluid and acts on each portion of the surface of the particles: this effect is quite simply the buoyant force. Dispersive force, on the other hand, only acts on the portion of contact between the particles in collision, and therefore has a different nature from the pressure in the fluid.

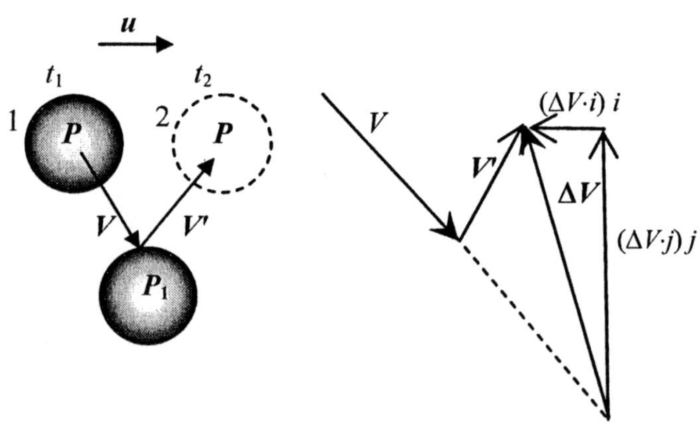

Figure 5.4: Diagram showing the dispersive force as a result of the collisions between the particles in motion [4].

As shown in Figure 5.5, when a particle collides with one positioned on the channel bed, it is subject to in the direction ΔV, and, according to the third principle of dynamics, P transmits the particles on the bed a force equal and opposite to the dispersive force, which is marked with F in Figure 5.5.

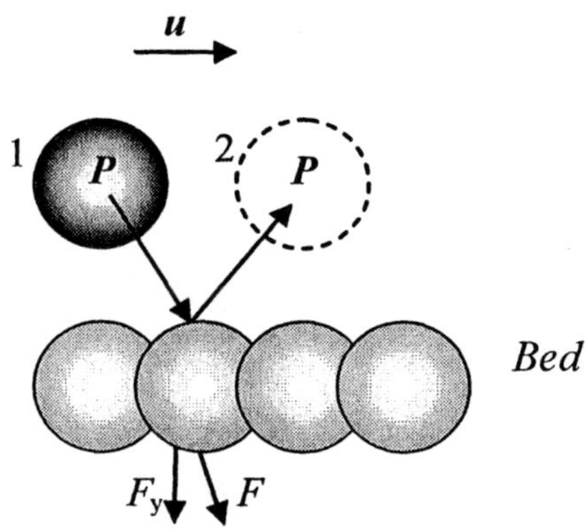

Figure 5.5: Reactionary force acting on the particles on the channel bed [4].

F is a reactionary force, in fact, the vertical component of F, F_y, plays a determining role in the stability of the bed: motion triggering of a bed of sediments is partly linked to lift (see section 4.3.1), and, to a certain extent, F_y counterbalances lift itself, thus preventing new particles from joining streamflow motion.

In the inertial region, Bagnold presented a simple analysis for estimating the constitutive relationship, by recalling the kinetic theory of gases and neglecting the surrounding fluid, in describing the interaction of spherical granules in a flow with a velocity $u(y)$; both the momentum exchange during a single collision and the frequency of collision are proportional to the relative velocity of the two layers, thus causing a quadratic dependency between stress and shear rate. The constitutive relationship discovered by Bagnold for normal stress and shear stress is as follows:

$$\tau_{yy} = \alpha\rho_s\lambda f(\lambda)d^2\cos\phi_d(\mathrm{d}u/\mathrm{d}y)^2 \tag{5.8}$$

$$\tau_{xy} = P\tan\phi_d \tag{5.9}$$

where ρ_s is grain density, a is a constant (0.042 in Bagnold's experiments), ϕ_d is the angle of dynamic friction, unknown, depending on the collisional conditions, $f(\lambda)$ is an unknown function of linear concentration λ, defined above. For the latter Bagnold found that $f=\lambda$ for $\lambda<14$ and that for $\lambda>14$ it increases very rapidly. Equation (5.8) has not only been confirmed by experiments (Savage and

McKeown [58] and Savage and Sayed [53] amongst others), but can also be connected in the ambit of kinetic theory and granular temperature as follows. From a dimensional point of view, in a model in which fluid interactions are neglected and the gravitational effect is purposely concealed, assuming liquid and solid of equal density, the only dimensional quantities involved are ρ_s, solid density, d, a characteristic dimension of the solid and γ, shear rate chosen as a timescale, for which $\tau = f(\rho_s, d, \gamma)$ and the principle of dimensional homogeneity applied to this functional dependence for the stresses, considering that $\left[\frac{M}{LT^2}\right] = \left[\frac{M}{L^3}\right]^1 \cdot [L]^2 \cdot \left[\frac{1}{T}\right]^2$, gives:

$$\tau_{yy} = \rho_s f(C) d^2 (\mathrm{d}u/\mathrm{d}y)^2 . \tag{5.10}$$

It is also interesting to note that, as granular temperature (see Section 3.5.1), has the dimension of a velocity squared, it must be proportional to $d^2\gamma^2$ for dimensional homogeneity. Finally, the viscosity of a material varies in a linearly proportional way to the shear rate and is therefore proportional to the square root of the granular temperature; in the same way that kinetic theory shows that the viscosity of a gas varies with the square root of the thermodynamic temperature. On the other hand, by assuming that no long radium forces act between the particles, the transport of any property must be on the chaotic motion of the particles, and the rate of this transport must be proportional to the entity of the fluctuation velocity, and therefore to the square root of the granular temperature [35]. It can be observed that the eqn (5.10) has a similar form to the theory of Prandtl's mixing length for the turbulence of fluids, thus indicating that the same mechanics that leads to Bagnold's dispersive stress is also responsible for the turbulence of the mixture. The turbulent model is expressed by:

$$\tau_t = \rho_f k_v^2 y^2 (\mathrm{d}u/\mathrm{d}y)^2 \tag{5.11}$$

and underlines the non–Newtonian nature of turbulent flows; in which ρ_f is fluid density, k_v is the von Karman constant (0.4 for pure water), which, in the case of water–sediment mixtures, depends on volumetric concentration [36] and y is the distance from the wall. Thus, as with turbulent transport, turbulence is responsible for sediment suspension, in a parallel way, in debris flows, dispersive pressure is mainly responsible for particle dispersion. In many granular flows both mechanisms are important to dispersion and should therefore be included in the model [15]. Despite the fact that Bagnold's theory contains certain empiricisms, his model has been used as the basic theory in developing constitutive relationships for rapid granular flows by Shen and Ackermann [46] and by Pasquarell et al. [50], amongst others. In reality, the use of Bagnold's formulae is debatable, because the results strongly depend on the conditions of his experiments, thus causing, in some cases, a number of contradictions, as Iverson illustrates [18]. We consider a solid–liquid mixture in a channel enclosed by parallel, vertical walls and guided by the force of gravity. Irrespective of mixture rheology, for symmetry's sake, the shear rate must become zero at the centre of the channel.

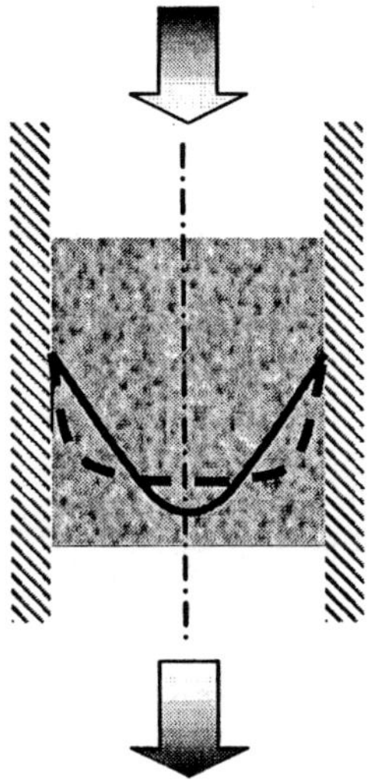

Figure 5.6: The flow of a solid–liquid mixture in a channel with vertical and parallel walls under the action of gravity; according to the granular temperature, various velocity profiles are possible, however, they must all satisfy the symmetry by which the shear rate must become zero in the centre [18].

The zeroing of the shear rate invalidates the use of Bagnold's equation, which for null rates also requires the cancellation of the shear rate at the centre of the channel, however, this contrasts with the presence of a body force such as gravity, which guides the flow.

Bagnold's experiments do not highlight this contradiction because the pressure gradient linked to gravity or other forces independent of the shear rate, are voluntarily ignored. The assumptions of his experiments purposely camouflage the effect of gravity in order to measure stress at certain given shear rates. However, in debris flows, gravity induces a stress that increases with depth and the shear rates follow the same pattern, and they, alone, cannot therefore constitute a valid model for debris-flows.

Takahashi [60] also directly applied Bagnold's equation to the analysis of a steady debris flow along an inclined channel, assuming that the Savage number tends to the infinite (effect of null friction), that there is no shear at the channel bed, that the grains are uniformly dispersed within the flow and that there is no excess of pressure in the interstitial fluid. Consequently, in order to balance the action of gravity, collisional stresses vary linearly in the depth of the streamflow. Takahashi's approach, however, excludes the presence of layers of nondeformed material, this not only contrasts with experimental observations and the viscoplastic model, but also with the observation that mass of highly compacted material due to the action of friction, must be present during the first and last stages of motion, when the debris material mobilises and deposits, respectively [18]. Takahashi's application of Bagnold's equations also involves the assumption of a greater value for the constant a (equal to 0.5). This difference in the orders could indicate that:

- other factors, such as the turbulent shear stress, which is also proportional to the square of the shear rate, may not have been accurately incorporated into the previous equation;
- Bagnold's results obtained for a flow ignoring force of gravity could not be directly applicable to a flow subject to gravitational action (such as a debris flow along an inclined plane);
- Takahashi [33] does not attribute this disagreement to the inadequacy of the equations of the dispersive model so much as to the neglected effect of the seepage of surface water into a channel bed that is not completely saturated.

5.4.3 Mixed frictional–collisional regime

We have already mentioned that the stresses within a granular material are transmitted by forces exerted at points of mutual contact between particles. When the individual particles are amply spaced within the flow, the single contacts are brief and the collision is dominant in determining momentum exchange. On the other hand, for slow deformation and high solid concentration, the contacts are semipermanent and the rubbing of the particles against one another plays a decisive role in momentum transport. The constitutive relationships available for these two boundary situations have been discussed previously; however, many cases of practical interest fall into the interval of these extremes, where collision and sliding friction are both significant [49]. Some researchers (McTigue, [52]; Johnson *et al.*, [56]) proposed constitutive relationships for granular materials shearing within the intermediate region, which consider both terms that do not depend on the deformation rate and on dependant ones with a flow behaviour index of 2, and therefore typically dilatant. For example, one can see the constitutive equation proposed by McTigue [52] for a flow of granular material along a slope:

$$\tau = \tau_c \cos\phi + \eta_1\left(C^2 - C_0^{\ 2}\right)\sin\phi + \eta_2\left(C_m^{\ 2} - C^2\right)\left(\frac{du}{dy}\right)^2, \tag{5.12}$$

where η_1 and η_2 are coefficients to be calculated, C_0 and C_m are the minimum and maximum solid concentration, respectively. The sum of the first two terms of the right-hand member of eqn (5.12) represents the yield stress τ_y that must be overcome in order for a flow to occur. A stress–strain relationship similar to eqn

(5.12), for a mixed friction-collisional regime flow can generally be expressed with the following equation:

$$\tau = \tau_y + \alpha \left(\frac{du}{dy} \right)^2 . \tag{5.13}$$

If τ_y and α are constant, the corresponding components of velocity in steady, uniform and bidimensional conditions are those shown in Table 5.3.

Table 5.3: Velocity components for a uniform and bidimensional debris flow in a friction-collision regime.

Velocity distribution	$u = \frac{2}{3}\sqrt{\frac{\rho_m g \sin\theta}{\alpha}}\left[H^{1.5} - (H-y)^{1.5}\right]$ for $0 \le y \le H$
Surface velocity	$u_s = \frac{2}{3}\sqrt{\frac{\rho_m g \sin\theta}{\alpha}} H^{1.5}$ for $H \le y \le h$
Velocity defect	$\frac{u_s - u}{u_s} = \left(1 - \frac{y}{H}\right)^{1.5}$ for $0 \le y \le H$
Average velocity	$U = \frac{2}{5}\sqrt{\frac{\rho_m g \sin\theta}{\alpha}} H^{1.5}\left(1 - \frac{2}{5}\cdot\frac{H}{h}\right) = \left(1 - \frac{2}{5}\cdot\frac{H}{h}\right) u_s$

As usual, θ is the bed slope, h is flow depth, u_s is the velocity at the free surface, U is mean velocity calculated by integrating on the flow depth and then dividing by its depth, ρ_m is mixture density, g gravitational acceleration and finally H is the height of flow depth at the point in which a shear stress equal to the yield stress is applied and $h - H = \tau_y / (\rho_m g \sin\theta)$ represents the thickness of the upper part in which relative velocity does not exist.

5.4.4 Macroviscous regime

It is thought that the existence of solid particles diluted in suspension in a biphasic system alters the flow of the fluid part of the mixture, thus creating an increase in viscous dissipation. Collision conditions are macroviscous [7], in that all the momentum acquired by a particle in a collision is transferred to the interstitial fluid prior to the subsequent collision, and this determines the dispersion of the grains throughout the entire depth of the flow. The relationship between stress and strain is similar to that for a Newtonian fluid, until the solid concentration exceeds approximately 9%. Therefore, in this regime, the fluid–granule mixture can be treated as a generalised Newtonian fluid, with apparent viscosity μ_a, which depends not only on the properties of the fluid and on temperature, but also on sediment concentration.

$$\tau = \mu_a \frac{du}{dy}. \tag{5.14}$$

One of the first expressions introduced for estimating the apparent viscosity of a suspension (see Section 3.3.4) was proposed by Einstein [37]: $\mu_a/\mu_w = (1 + 2.5C)$, where μ_w is the viscosity of water and C is, as usual, the solid concentration by volume. As Einstein's equation, and its extensions, are only valid for very dilute spherical suspensions with negligible interaction between particles, a suitable correction could be considered for application to a more concentrated suspension, with significant interaction between particles. By adding to the sediment concentration the effect of the thin layer of water that adheres to the surface of the particles, Chu [61] obtains an extension of Einstein's equation: $\mu_a/\mu_w = (1 + 2.5KC)$, where $K = (V_s + V_w)/V_s$ is the ratio between the sum of the volume of the sediments and the surrounding water and the volume of the sediments and it is slightly higher than the unity; if we then subsequently also consider the variation in the porous space due to the collision of the particles, Chu concluded that $\mu_a/\mu_w (1 + \varepsilon KC)^{-2.5}$, where ε is the coefficient that indicates the increase in the intergranular space and is equal to 4.5/π (approximately 1.43) for uniform spheres. Krieger and Dougherty [62] obtained $\mu_a/\mu_w = (1 - KC)^{-B/K}$, with B close to 2.5, Bagnold [7], in his experiments with coaxial cylinders in a macroviscous regime, used $\mu_a = 2.25\lambda^{1.5}\mu_w$. If the apparent viscosity μ_a is constant, the velocity components for a steady, bidimensional and uniform debris flow can be easily obtained from eqn (5.14) and are shown in Table 5.4; as usual, θ is the bed slope, h is the flow depth, ρ_m is mixture density, g is the gravitational acceleration, u_s is the velocity at the free surface and U is the mean velocity calculated by integrating on the depth of the flow and then dividing by depth itself.

Table 5.4: Velocity components for a uniform and bidimensional debris flow in a macroviscous regime.

Velocity distribution	$u = \frac{\rho_m g h \sin\theta}{\mu_a}\left[\frac{y}{h} - \frac{1}{2}\left(\frac{y}{h}\right)^2\right]$
Surface velocity	$u_s = \frac{1}{2}\frac{\rho_m g h^2 \sin\theta}{\mu_a}$
Velocity defect	$\frac{u_s - u}{u_s} = \left(1 - \frac{y}{h}\right)^2$
Average velocity	$U = \frac{1}{3}\frac{\rho_m g h^2 \sin\theta}{\mu_a} = \frac{2}{3}u_s$

5.4.4.1 Takahashi's viscous model

As previously stated, a viscous debris flow contains mainly cohesive and fine particles, as the interstitial fluid is particularly dense due to its clay and silt composition. Therefore, the prevalent stresses in the dynamics of the flow are viscous ones rather than collisional and/or turbulent ones, and there is no rigid plug in the flow, in fact, a vertical shear rate exists up to the surface [41]. This type of debris-flow has the following characteristics: greater particle diameter, between 0.1 mm and 10 cm, solid concentration by volume (see Section 2.2) between 0.53 and 0.8. In the model proposed by Takahashi *et al.* [63] the mechanisms responsible for the dispersion of the coarser particles within the viscous medium, seem to be the moving closer of the particles on planes adjacent in relative motion, and a lower mobility of them towards the (lower) regions with higher concentrations. Within the highly viscous interstitial fluid, the former causes a runoff that moves the particles perpendicular to the main direction under a dispersive force that can be attributed to the action of the drag force. The latter occurs if the distribution of the coarser particles is not isotropic; this causes a variable apparent viscosity that is higher in the region with the highest concentration, consequentially, resistance to migration towards this area will be higher [63]. The motion equation in the model proposed by Takahashi et al. [63] is as follows:

$$\int_y^h \left[C\rho_s + (1-C)\rho_f\right] g \sin\theta \, dy = \mu_a \gamma , \tag{5.15}$$

where μ_a is the apparent viscosity of the mixture, C and ρ_s the concentration by volume and sediment density, ρ_f is fluid density, g is gravitational acceleration, θ bed slope, h is streamflow depth and $\gamma=(du/dy)$ shear rate in the y direction perpendicular to flow. If we assume a mean particle concentration $\overline{C}$, we can integrate the previous equation with the change of variables $Y=y/h$ and $u' = u/\sqrt{gh}$ and with the boundary condition u' $(Y=0)=0$, thus obtaining a velocity distribution near to the typical parabolic profile of the laminar flow (see Figure 5.7):

$$u' = \frac{\rho_f \sqrt{gh}h \sin\theta}{\mu_a}\left(1 + \frac{\rho_s - \rho_f}{\rho_f}\overline{C}\right)\left(Y - \frac{Y^2}{2}\right). \tag{5.16}$$

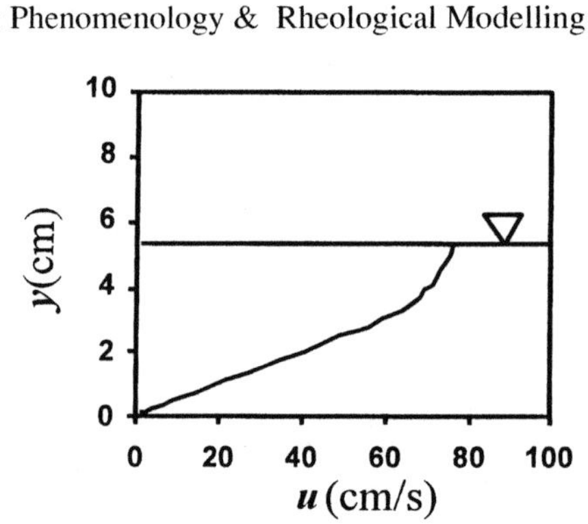

Figure 5.7: Theoretical velocity distribution in Takahashi's viscous model (θ=12°).

If we then assume that concentration is not uniform in the y direction, perpendicular to the direction of the main flow, we must introduce a diffusion equation; that proposed in Takahashi's model is a direct consequence of the mechanisms held responsible for the dispersion of the coarser particles within the viscous medium. The proposed diffusion equation is as follows:

$$\frac{\partial C}{\partial t} = -\nabla \cdot \left(N_{c} + N_{\mu_a} + N_{s}\right), \tag{5.17}$$

in which N_c relates to the flow of particles perpendicular to the direction of the main flow and due to the coming closer of the particles on two adjacent planes and in relative motion to one another within the viscous medium; N_{μ_a} relates to the flow due to the spatial variation of the apparent viscosity of the mixture and N_s relates to the gravitational flow due to the particles' tendency to deposit under their own weight.

$$N_{c} = -K_{c} d^{2}\left(C^{2}\nabla\gamma + C\gamma\nabla C\right) \tag{5.18}$$

$$N_{\mu_a} = -\frac{K_{\mu_a}\gamma C^{2} d^{2}}{\mu_a}\frac{\mathrm{d}\mu_a}{\mathrm{d}C}\nabla C \tag{5.19}$$

$$N_{s} = -\frac{2}{9}C\frac{d^{2}\left(\rho_{s} - \rho_{f}\right)g\cos\theta}{\mu}\left(\frac{1-C}{\mu_{a}/\mu}\right), \tag{5.20}$$

in which, in addition to the usual factors, we also have K_c and K_{μ_a}, which are suitable constants, d, which is the particle diameter and μ which is the viscosity of the interstitial fluid. With a hypothesis of steadiness and assuming the previous expressions for the flows, one obtains a concentration distribution (see [63]) that is independent of particle diameter.

5.4.5 Viscoplastic regime

The viscoplastic assumption for a debris flow is widely reliant on the idea that a high concentration of sediment in the flow increases viscosity and gives resistance to shear, due to the action of an interstitial fluid with a finely granular structure that produces cohesion and due to the action of the coarser particles, which are subject to internal friction. Viscoplastic materials have a finite yield stress and run off as Newtonian fluids when it is exceeded; recalling Bingham's plastic models (see Section 2.3.1.2), the constitutive equation becomes:

$$\tau = \tau_y + \mu_b \frac{du}{dy} \quad \text{for } \tau > \tau_y$$

$$\frac{du}{dy} = 0 \quad \text{for } \tau < \tau_y, \tag{5.21}$$

which is a linear relationship between stress and strain, in which τ_y is Bingham's yields stress (or shear resistance) and μ_b Bingham's viscosity. The debris flows that show a yield stress generally drive to the appearance of a portion, which looks "freezed", called a rigid plug, in each region where shear stress is lower than yield stress. Bingham's models can be generalised in such a way that yield stress depends on Coulombic friction and viscosity on the shear rate, although, in general, in applications with a steady and uniform flow these parameters are assumed to be constant. In Bingham's model, yield stress is equal to Coulombic stress, if one assumes that the internal friction angle is null, so that cohesion alone controls the failure. This model has been used to describe mudflows and hyperconcentrated streamflows (O'Brien and Julien, [15]; Fei, [22]), and is particularly common in Chinese literature. Viscoplastic models that are based on the power law, such as that of Hershel–Bulkley (Coussot, [16]; Bin and Huilin, [25]), are herein neglected as they are purely empirical models that can only be obtained by means of rheometric tests on the material (see Section 2.3.2).

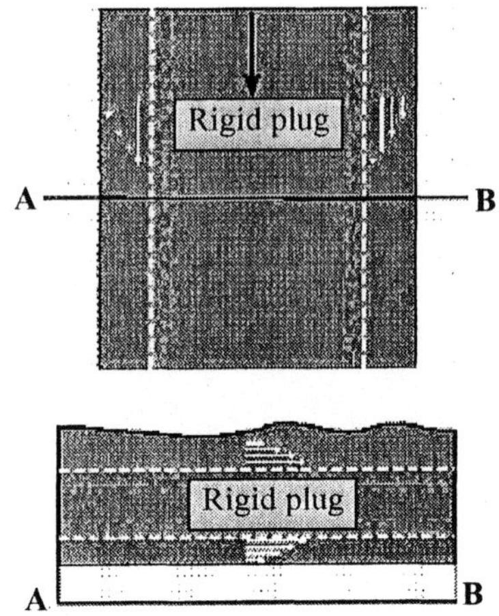

Figure 5.8: Debris flows with yield stresses usually lead to the formation of a rigid plug in each region where shear stress is lower than yield stress.

By using laboratory experiments conducted on steady and uniform flows in circular channels and observations made on natural debris flows, Johnson [9] divided yield stress τ_y into two parts, one due to Coulombic friction, the other to cohesion between grains; assuming that the first contribution for a debris-flow is proportional to normal stress σ acting on shear planes, in the same way as the Mohr–Coulomb criterion (eqn (5.3)), he developed one of the first models for debris-flows, known as Coulomb's viscous model and described by the following constitutive equation:

$$\tau = \tau_c + \sigma \tan \phi + \mu_b \frac{du}{dy}. \qquad (5.22)$$

The model describes the behaviour of a one-phase material that, following the exceeding of the plastic yield stress, flows as a viscous liquid. The first two terms of the second member represent this yield stress, $\tau_y = (\tau_c + \sigma \tan \phi)$, which must be exceeded in order for motion to take place. It can be observed that, as normal stress is generally dependent on shear rate, the addition of the last term is, in some ways, superfluous [1]. Bingham's model and Coulomb's viscous model are generally known as viscoplastic models. The values of the rheological parameters τ_y and μ_b are not constant but depend largely on the properties of the fluid–granular mixture [20] such as: solid concentration, type of clay and silt, shape and the dimensional distribution of particles, temperature and electrochemical properties of the liquid component. For example, an experimental study performed on finely granular suspensions and with a high (between 0.44 and 0.66) sediment concentration by volume (page 17) carried out by Major and Pierson [64], shows that yield stress and viscosity undergo a variation in order when sediment concentration oscillates even slightly, between 2% and 4%. Viscoplastic models are easily applicable and can explain certain key debris-flow characteristics, such as the existence of a rigid plug composed of relatively nondeformed material, that runs along the central part of the channel, and the suspension of coarser particles that fluctuate in the interstitial fluid [3]. The basic assumption in viscoplastic models is that the interstitial fluid is responsible for both yield stress and the mixture's viscous behaviour, and therefore the properties of the interstitial fluid necessarily cause the dynamic resistance to motion, as shown by eqns (5.21) and (5.22), without, however, implicitly including the interparticle effects and fluid–particle interactions. The main limits to Bingham's model in application to debris-flows are as follows [3]:

- They simulate the rheology of the fine and liquid fraction of debris mixtures only, whereas data fields show that the interactions between coarser solid grains and the interstitial fluid substantially alter its overall behaviour.
- They assume that momentum transport and energy dissipation take place due to viscous effects only. They neglect dissipation not linked to the shear rate that occurs by means of mutual contact between granules or on the boundary.
- They only represent one boundary behaviour, which occurs when the Bagnold's number is null and Savage's number tends towards the infinite, which is typical in creep and slow landslides.
- They use the adherence condition; which means that the material leaves behind a continuous layer of deposited sediments, but in reality, these deposits have not been observed, whereas the grains could slide, collide and roll on the bed.
- It should also be pointed out that viscoplastic and macroviscous models describe the laminar flow of muddy residues only; in the case of turbulent flows, turbulent stress must be considered proportional to the square of the shear rate.

Table 5.5 shows the velocity components for a steady, uniform and bidimensional debris flow when yield stress τ_y and viscosity μ_b are constant; θ is the bed slope, ρ_m is the mixture density, g is the gravitational acceleration, h is the flow depth, H is the height from the flow base at the point in which a shear stress equal to yield stress is applied, u_s is the velocity of the free surface and U is the mean velocity calculated by integrating on the flow depth and then dividing by depth itself.

Table 5.5: Velocity components for a steady, uniform and bidimensional debris flow in the viscoplastic regime.

Velocity distribution	$u = \dfrac{\rho_m g H^2 \sin\theta}{\mu_b}\left[\dfrac{y}{H} - \dfrac{1}{2}\left(\dfrac{y}{H}\right)^2\right]$	for $0 \le y \le H$
Surface velocity	$u_s = \dfrac{\rho_m g H^2 \sin\theta}{2\mu_b}$	for $H \le y \le h$
Velocity defect	$\dfrac{u_s - u}{u_s} = \left(1 - \dfrac{y}{H}\right)^2$	for $0 \le y \le H$
Average velocity	$U = \dfrac{\rho_m g H^2 \sin\theta}{\mu_b}\left(\dfrac{1}{2} - \dfrac{1}{6}\dfrac{H}{h}\right) = \left(1 - \dfrac{1}{3}\dfrac{H}{h}\right)u_s$	

5.4.6 Viscoplastic-collisional regime

In contrast to mixtures composed entirely of muddy material, coarser mixtures have a wider distribution and higher sediment concentration, which are characteristic of a viscoplastic-collisional regime [49]. In fact, in this case the interactions of coarser particles and fluid viscosity could play a significant role in momentum exchange.

5.4.6.1 O'Brien and Julien's quadratic model

On the one hand, at high deformation rates, in other words, in the granular-inertial regime, the dilatant model assumes that the stresses depend on the square of the shear rate due to the effect of collisions and friction; on the other, the observations, for low deformation rates, show that shear stress, above the yield stress, is proportional to the shear rate. O'Brien and Julien [15, 28] proposed an intermediate quadratic model that not only satisfies extreme conditions, but also the transition region and that includes cohesion, viscous fluid–sediment interactions, collisions and turbulence as dominant effects.

$$\tau = \tau_y + \mu_d \frac{du}{dy} + \zeta\left(\frac{du}{dy}\right)^2 \quad \text{or:} \tag{5.23}$$

$$\tau = \tau_y + \mu_d \frac{du}{dy} + \left(\mu_c + \mu_t\right)\left(\frac{du}{dy}\right)^2 . \tag{5.24}$$

In the equation above, μ_d is the dynamic viscosity of the fluid interacting with the sediment particles and τ_y is the yield stress, $\tau_y = \tau_c + p_s \tan\phi$, which describes both the stress τ_c, due to the cohesion between the fine particles and that is assumed to be a property of the material and independent of the deformation rate and Mohr–Coulomb's stress, which depends on intergranular pressure p_s and the material's angle of repose ϕ. Viscosity and yield stress show an exponential trend as a function of the sediments' volumetric concentration (see Section 2.3.2.1). Moreover, $\zeta = \mu_c + \mu_t$ is the inertial stress parameter, which combines the dispersive and turbulent effects induced by the collisions between particles, both proportional to the square of the shear rate. The two effects are combined because at high concentrations of coarser particles, the dispersive stress is dominant, whereas for high finer sediment concentrations, yield stress and viscous stress prevail over turbulent stress. Moreover, μ_c is the dispersion parameter defined by Bagnold as $\mu_c = a\rho_s\lambda^2 d^2$ (see Section 5.4.2.1), in which a=0.01 [32], and lastly, μ_t is the turbulence parameter, with $\mu_t = \rho_m l_m^{\,2}$ in which ρ_m and l_m are the density and mixing length of the mixture. It is assumed that $l_m = ky$, is a function of the distance y from the bed, with k Von Karman's constant (0.4 for pure water); the turbulence parameter is generally far lower than the dispersion parameter [65].

Using Bagnold's data [7] and that of Savage and Mckeown [58] the authors adopt eqn (5.23) in order to obtain an estimate of τ_y, μ_d and ζ for a number of debris-flow deposits (Figure 5.9).

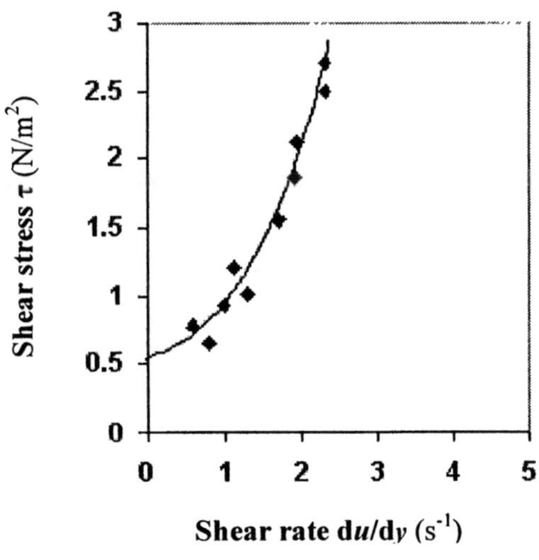

Figure 5.9: The graph shows the results (squares) of the laboratory experiments conducted by O'Brien and Julien [15] using a rotating viscosimeter to test the stress–strain relationship of a number of debris-flow deposits and the parabolic relationship obtained from regressional analysis on the same results, of which the expression is as follows: τ=0.52+0.031($\partial u/\partial y$)+0.31($\partial u/\partial y$)^2.

By substituting the expressions above for μ_c and μ_t and dividing each side of eqn (5.24) by $\mu_d \cdot (du/dy)$, one obtains a dimensionless formulation of the model:

$$\frac{\tau - \tau_y}{\mu_d \dfrac{du}{dy}} = 1 + a\left(\frac{\rho_s \lambda^2 d^2}{\mu_d}\frac{du}{dy}\right)\left[1 + \left(\frac{\rho_m l_m^2}{a\rho_s \lambda^2 d^2}\right)\right]. \tag{5.25}$$

By introducing the following dimensionless ratio, eqn (5.25) can be reformulated using the eqn (5.26):

$$\tau^* = 1 + a\left(1 + T_d^*\right)D_v^*, \tag{5.26}$$

in which:

- $\tau^* = \dfrac{\tau - \tau_y}{\mu_d \dfrac{du}{dy}}$ is a dimensionless shear stress which, for values close to zero, indicates a rigid behaviour typical of a solid, for values close to the unit, indicates a Newtonian (τ_y=0) or Bingham ($\tau_y \neq 0$) behaviour and for values notably higher than the unit indicates the prevalence of dispersive and turbulent stresses.
- $D_v^* = \left(\dfrac{\rho_s \lambda^2 d^2}{\mu_d}\dfrac{du}{dy}\right)$ is the dimensionless ratio between the dispersive and viscous effect, which, for high values, indicates the dominance of stresses that are caused by intergranular contacts, whereas for low values viscous dissipation prevails. This parameter increases with increases in granule dimension.
- $T_d^* = \left(\dfrac{\rho_m l_m^2}{a\rho_s \lambda^2 d^2}\right)$ is the dimensionless ratio between the turbulent and dispersive effect; at strong solid concentrations the dispersive stresses are dominant, whereas at low solid concentrations turbulent and dispersive stresses prevail.

In eqn (5.26) only three dimensionless parameters appear; the second term of the right member represents the deviation from Bingham's model and both Newtonian and non-Newtonian fluids can be characterised by this equation as a function of the three above-mentioned parameters. Moreover, by assuming the physical properties of the mixture to be constant, eqn (5.26) represents a linear function of τ^* and D^*_v. Once again, by using the data of Bagnold [7] and that of Savage and Mckeown [58] to estimate the overall coefficient that appears in the dimensionless and linear formulation of the quadratic model, the authors found that $a\left(1+T_{d}^{*}\right)=0.0087$; Figure 5.10 shows τ^* as a function of D^*_v. One can also note that the value of the overall coefficient is very close to the value of the parameter a (a=0.01) found by Bagnold when $T^*_d \to 0$, that is when turbulent forces are negligible; moreover, with equal shear rates and mixture concentration, D^*_v is greater for granular mixtures, which thus justifies the adoption of dilatant models for this type of mixture. One proceeds in this way, bypassing the parameter T^*_d, because there is currently no laboratory data to evaluate the mixing length of the mixture.

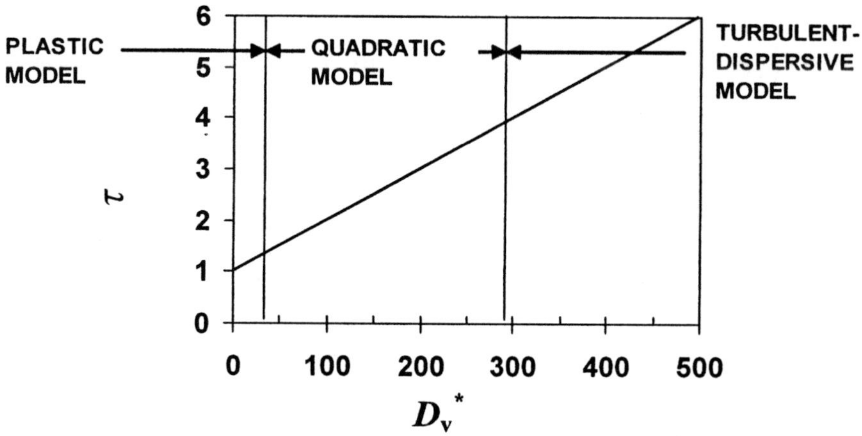

Figure 5.10: Representation of eqn (5.26) and definition of application field of the Bingham's plastic model, the quadratic model and the dispersive-turbulent model [65].

When the dimensionless parameter D_v^* is less than 30, the model corresponds to the simple case of Bingham's plastic model (τ^*=1); whereas if D_v^* is greater than 400 it corresponds to the purely collisional model, as in inertial regimes dispersive and turbulent stresses prevail, lastly, for intermediate values it is possible to refer to the quadratic model which, foreseeing the simultaneous presence of a number of aspects, has a wider generality. At the present time, the quadratic model could be the best way of tackling debris flows, although it is not the most convenient formula from an applicative point of view [49]. However, there are contrasting opinions on the validity of the quadratic model, for example, according to Iverson [3], the linear sum with which viscoplastic and dilatant approaches are combined, would not resolve the inefficiencies that have been previously demonstrated for these models.

Lastly, the parametric formulation of the quadratic model can be used to classify hyperconcentrated streamflows on the basis of the relative prevalence of the stresses acting in the mixture:

- mudfloods for $D_v^* > 400$ and $T_d^* > 1$;
- mudflows for $D_v^* < 30$;
- debris flows for $D_v^* > 400$ and $T_d^* < 1$.

Despite the fact that the criterion proposed by O'Brien and Julien [15] has been applied to experimental rheometric tests only, it can give useful indications for the rheological characterisation of solid–liquid flows; in particular, the application of this criterion to experimental flows in the laboratory and in the field [66] has shown that with an increase in the inclination of the channel beds in which solid–liquid mixtures run off, one can observe a gradual passage from flows characterised by the prevalence of turbulent-type stresses (which can be classified as water flows with or without solid transport) to flows more correctly classifiable as debris flows.

5.4.6.2 Chen's generalised viscoplastic model

Chen [13] attempted to introduce a general form of the constitutive equations for debris flows by starting with a generalised approach of the dilatant model proposed by Bagnold [7] and extending it to the viscoplastic-collisional regime. The analysis Chen conducted on the various flow regimes of a granular mixture identifies, in parallel with Bagnold's study, three states: a quasistatic one, which is a condition of incipient movement with practically plastic behaviour, a macroviscous one at low shear rates, in which viscosity determines mixture behaviour, and finally a granular inertial state, typical of rapid flowing granular mixtures dominated by intergranular interactions. Chen [13] developes his study starting from the assumption that a general solution should be applicable through all three regimes.

Chen [20] observes that an applicable model able to realistically describe debris flows should consider both an independent term and one dependent on the shear rate in a perpendicular direction to the main flow (Figure 5.11), as well as the main rheological properties: the effect of normal stress (Weissenberg effect, given by the second member on the right in eqn (5.28) below) and the Mohr–Coulomb yield criterion.

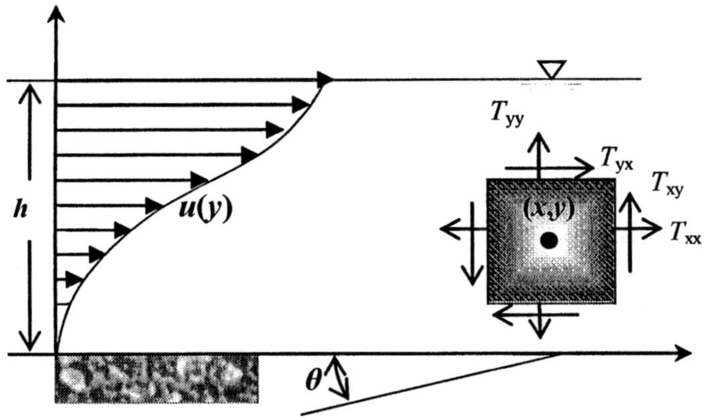

Figure 5.11: Definition diagram for a uniform and steady debris flow under a bidimensional state of stress in Chen's generalised viscoplastic model.

The importance of introducing a term independent of the shear rate, a term that is not included in the original formulation of Bagnold's dilatant model, lies in the fact that when the granular mixture passes from a dynamic state to a quasistatic condition, the total stress is exclusively supported by quasistatic stress, in full agreement with the principles of soil mechanics, and the angle of dynamic friction ϕ_d of the dilatant model becomes equal to the angle of static friction. By introducing the hypothesis according to which the viscous effect should not coexist with turbulent and dispersive effects, Chen [13] developed a generalised model for a viscoplastic fluid (GVF model) in the following way:

$$T_{yx}\left(=T_{xy}\right)=c\cos\phi+p\sin\phi+\mu_1\left(\frac{\mathrm{d}u}{\mathrm{d}y}\right)^{\eta} \tag{5.27}$$

$$T_{yy}\left(=T_{xx}\right)=-p+\mu_2\left(\frac{\mathrm{d}u}{\mathrm{d}y}\right)^{\eta}. \tag{5.28}$$

The equations given above describe the shear and normal stress at the point (x,y); in them c is cohesion (force of adhesion that acts between the particles per surface unit), ϕ is the angle of internal friction, p is the pressure, μ_1 and μ_2 are the consistency indexes, η is the flow behaviour index, which varies between 1 and 2 when passing from a macroviscous to an inertial regime. By following the Mohr–Coulomb criterion, one can introduce a yield stress, by assuming $s = c\cos\phi + p\sin\phi$. When passing from a quasistatic regime to a fully dynamic one, the values of c and ϕ become negligible. Thus Chen's model can cover the entire spectrum of Newtonian models, viscoplastic models (η =1), dilatant models (η =2), and the power law (η varies), according to how yield stress, consistency and flow behaviour indexes are chosen. However, the usefulness of the equations above is closely connected to the possibility of accurately determining the coefficients that appear, in other words τ_c, ϕ, p, μ_1 and η, the latter being especially difficult to calculate given the close dependency on flow regime.

For homogeneous materials, it is possible to assume concentration C and the consistency indexes μ_1 and μ_2 are constant. By indicating with $\rho_m = C(\rho_s - \rho_f) + \rho_f$ the mixture density, it is possible to express the constitutive equations as follows:

$$\rho_m g(h-y)\sin\theta = c\cos\phi + p\sin\phi + \mu_1\left(\frac{du}{dy}\right)^\eta \tag{5.29}$$

$$-\rho_m g(h-y)\cos\theta = -p + \mu_2\left(\frac{du}{dy}\right)^\eta . \tag{5.30}$$

After various passages and with the hypothesis $u_{(y=0)}=0$ at the bed, one therefore obtains p and the velocity profile u:

$$p = \rho_m g(h-y)\cos\theta - \frac{\mu_2}{\mu_1}\rho^* g(H-y)\sin\theta \tag{5.31}$$

$$u = \frac{\eta}{\eta+1}\left(\frac{\rho^* g\sin\theta}{\mu_1}\right)^{\frac{1}{\eta}} H^{\frac{\eta+1}{\eta}}\left[1-\left(1-\frac{y}{H}\right)^{\frac{\eta+1}{\eta}}\right] \quad \text{for } 0\le y\le H \tag{5.32}$$

$$u = \frac{\eta}{\eta+1}\left(\frac{\rho^* g\sin\theta}{\mu_1}\right)^{\frac{1}{\eta}} H^{\frac{\eta+1}{\eta}} \quad \text{for } H\le y\le h . \tag{5.33}$$

In the equations above ρ^* is a constant defined by $\rho^* = \left[\frac{1-\text{ctn}\theta\sin\phi}{1+(\mu_2/\mu_1)\sin\phi}\rho_m\right]$ and H is that vertical coordinate for which normal and shear stress only present the term independent of the shear rate; in conclusion $(h-H)$ represents the thickness of the rigid plug. In a quasistatic condition, the granular mixture moves as a very slow landslide, whereas in fully developed dynamic conditions, the thickness of the rigid plug progressively diminishes ($H/h\rightarrow 1$).

For nonhomogeneous materials [67] concentration varies with flow depth; thus by deriving eqns (5.29) and (5.30), and substituting an adequate relationship between μ_1 and μ_2 and C, one obtains:

$$-\rho_m g\sin\theta = \frac{dp}{dy}\sin\phi + A_1\frac{d}{dy}\left[(1-KC)^{-B/K}\left(\frac{du}{dy}\right)^\eta\right] \tag{5.34}$$

$$\rho_{\mathrm{m}} g \cos\theta = -\frac{\mathrm{d}p}{\mathrm{d}y} + A_2 \frac{\mathrm{d}}{\mathrm{d}y}\left[(1-KC)^{-B/K}\left(\frac{\mathrm{d}u}{\mathrm{d}y}\right)^{\eta}\right], \tag{5.35}$$

with:

$$\mu_1 = A_1(1-KC)^{-B/K} \tag{5.36}$$

$$\mu_2 = A_2(1-KC)^{-B/K}, \tag{5.37}$$

A_1 and A_2 are numerical constants, K is a constant that describes the filling of the spaces in the mixture and B is Einstein's constant, which is equal to 2.50 for rigid spheres; the previous expressions for the consistency indexes are derived from those of Kreiger and Dougherty (see Section 5.4.4). Relative viscosity $\mu_{\mathrm{r}} = (1-KC)^{-B/K}$, which is given by the ratio between the viscosity of the granular mixture and the viscosity of water, herein plays a similar role to that of linear concentration λ of Bagnold's model, however, thanks to the parameter B/K, it is also possible to incorporate the macroscopic behaviour of the scattered particles. The value of B is slightly higher than that indicated by Einstein (B=2.50) and can vary from 2.5 to 4.5.

The resolution of eqns (5.34) and (5.35) requires the introduction of a relationship between density and pressure, for which a general form with constant temperature is given by:

$$\kappa = \frac{1}{\rho_{\mathrm{m}}^{\ r}} \frac{\mathrm{d}\rho_{\mathrm{m}}}{\mathrm{d}p}. \tag{5.38}$$

In eqn (5.38) k is the mixture compressibility, $\mathrm{d}\rho_{\mathrm{m}} = (\rho_{\mathrm{s}} - \rho_{\mathrm{f}})\mathrm{d}C$ and r is a suitable exponent; as density therefore depends on concentration, by varying k and r eqn (5.38) provides a wide range of possible relationships between the two variables, one of which was proposed by Mc Tigue [52]. By considering a dry granular mixture, composed of grains with density ρ_{s} and voids, whose mass is neglected in relation to that of the grains, mixture density is given by $\rho_{\mathrm{m}} = C\rho_{\mathrm{s}}$, C being solid concentration by volume of the grains (ratio between grain volume and mixture volume), by posing r=−1, eqn (5.38), with the previous assumptions, is altered to $\mathrm{d}p = \frac{\rho_{\mathrm{s}}^{2} \cdot C}{\kappa}\mathrm{d}C$; by integrating between a generic height y of the flow (Figure 5.11) and the free surface, where pressure becomes zero (having taken atmospheric pressure as a reference) and solid concentration by volume assumes a value indicated by C_{s}, one obtains the pressure–solid concentration relationship proposed by McTigue [52]:

$$p = \alpha\left(C^2 - C_{\mathrm{s}}^{2}\right), \tag{5.39}$$

where $\alpha = \rho_s^{2}/(2k)$; eqn (5.39) could also be extended to the case of water–sediment mixtures. Therefore by using eqn (5.38) one can obtain pressure and velocity profiles even for non homogeneous mixtures (see Chen, [67]).

Despite Chen's attempts to determine a generalisation of debris-flow modelling and to define the various parameters of the model, according to Iverson [3] even Chen's generalised model would constitute a formulation with numerous parameters, without, however, concretely incorporating the physical requisites described for debris-flows, which in any case remain latent behind mathematical modelling.

For the sake of completeness, Table 5.6 shows the velocity components for a steady, uniform and bidimensional debris flow in the generalised viscoplastic model according to the reference system in Figure 5.2, when the relative coefficients (μ_1 inconsistency index and η flow-behaviour index) are constant; θ is the bed slope, ρ_m is the mixture density, g is the gravitational acceleration, h is the flow depth, H is the height from the base of the flow to the point in which a shear stress equal to the yield stress is applied, u_s is the velocity at the free surface, U is the mean velocity calculated by integrating on the flow depth and dividing by the overall depth and, lastly, ρ^* is the constant previously defined.

Table 5.6: Velocity components for a uniform and bidimensional debris flow in a viscoplastic-collisional regime according to Chen's GVF model.

Velocity distribution	$u = \frac{\eta}{\eta+1}\left(\frac{\rho^* g \sin\theta\, H^{\eta+1}}{\mu_1}\right)^{\frac{1}{\eta}}\left[1-\left(1-\frac{y}{H}\right)^{\frac{\eta+1}{\eta}}\right]$ for $0 \le y \le H$
Surface velocity	$u = \frac{\eta}{\eta+1}\left(\frac{\rho^* g \sin\theta\, H^{\eta+1}}{\mu_1}\right)^{\frac{1}{\eta}}$ for $H \le y \le h$
Velocity defect	$\frac{u_s - u}{u_s} = \left(1-\frac{y}{H}\right)^{\frac{\eta+1}{\eta}}$
Average velocity	$U = \frac{\eta}{\eta+1}\left(\frac{\rho^* g \sin\theta\, H^{\eta+1}}{\mu_1}\right)^{\frac{1}{\eta}}\left(1-\frac{\eta}{2\eta+1}\frac{H}{h}\right) = \left(1-\frac{\eta}{2\eta+1}\frac{H}{h}\right)u_s$

5.4.7 Multilayer models

So far we have reviewed various expressions for debris-flow velocity (from Table 5.2 to Table 5.6) by adopting different models and in the hypothesis of uniform mixtures. The velocity-distribution profiles thus obtained have a parabolic trend; however, in many experimental tests [4, 14, 46, 55] "S" reversed shaped trends have been obtained, where the maximum shear rate is not obtained near the bed, but rather between the bed and the free surface. The main discrepancy is derived from the assumption of a debris-flow as a uniform mixture; in fact, the solid-concentration distribution is usually nonuniform due to the action of gravity and the lower layer of the flow could consequently have a higher concentration than the upper layer. Higher concentration means higher cohesion, friction and viscosity in the flow; for this reason, the velocity profile for nonuniform concentration is altered. The coefficients of the above-mentioned models closely depend on concentration and its distribution; and as concentration varies with flow depth, these coefficients are not constant, but rather vary with depth. As an alternative to this kind of models, Su et al. [68] proposed a multi layer model in order to simulate the velocity profile in steady and uniform flow conditions, by dividing the flow into two layers; one is the macroviscous layer near the bed and above this there is the inertial layer where the dispersive pressure and turbulent stress are equally important. Following comparisons with experiments they found that the multilayer model behaved better in simulating velocity distribution. The experiments conducted by Davies [42] however, show that the shear rate du/dy is greater near the bed, with a mean shear rate less than $10\ s^{-1}$; this could mean that flow at the bed tends to be inertial, whereas the upper layers are macroviscous: this assumption disagrees with the model proposed by Su et al. [68], moreover, bed deposits do not suggest different layers. According to Jan and Shen [49] however, we must observe that bed deposition could indicate whether sediment grading takes place or not (Section 3.4.4), however, it will never be able to demonstrate whether sediment-concentration distribution is actually uniform or not in the direction orthogonal to the direction of the main flow.

Independently of Su *et al.*, Wan [4] also proposed a multilayer model known as the laminated layer model, by dividing the debris flow into three regions from the bed to the free surface: a bed layer, in which an additional shear stress (which results from the contact of the particles and fluid viscosity) is dominant in momentum exchange; an inertial layer, where the dispersive stress of the grains is dominant and an upper viscoplastic layer, which can be represented by Bingham's model. According to the author, the velocity profile obtained using this model is very close to experimental results.

These two examples illustrate how much work is still to be done in this field.

5.4.7.1 Takahashi's multilayer model

In an inertial debris flow where the fluid contains a negligible quantity of fine cohesive material, fluid deformation exerts a negligible effect, whilst stresses linked to the brief and inelastic collisions of the coarser particles and the internal friction, caused by prolonged contacts between granules, could play a significant role: in other words a friction and collisional regime.

Takahashi *et al.* [33, 69] propose a so-called unified model of inertial debris flows, by altering the constitutive equations from the theory of granular mixtures and suspended load transport.

In this model, Takahashi hypothesises that the flow is composed of a lower layer, dominated by collisions, and an upper one composed of turbulent suspension. The relative extension of the two layers depends on the concentration and diameter of the particles; the range goes from *stony debris flows* in which only a collisional layer exists, to *muddy debris-flows* in which the entire flow is composed of a turbulent layer.

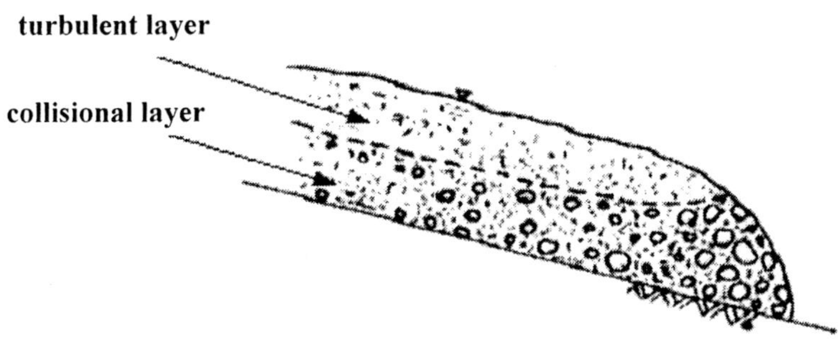

Figure 5.12: Diagram showing an inertial debris flow in the multilayer model proposed by Takahashi [33].

In the lower layer the particles move parallel to the bed with frequent collisions.

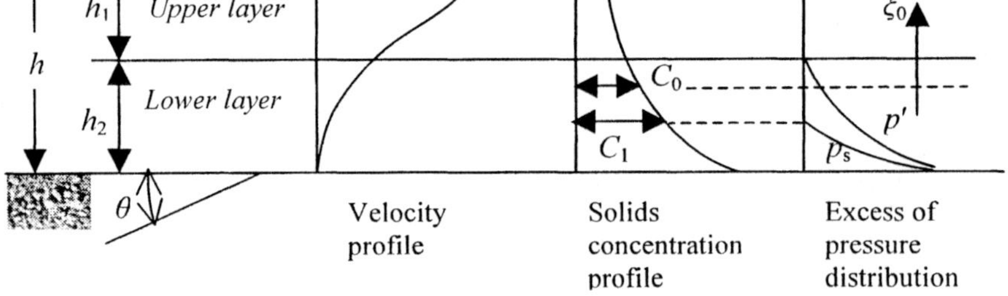

Figure 5.13: Definition diagram for an inertial debris flow [33].

For the lower layer, the stress-balance equations, in the main flow direction and perpendicular to it, are as follows :

$$\tau_c + \tau_k + \tau_s + \tau_t = p' \tan\theta + \rho g(h - y)\sin\theta \tag{5.40}$$

$$p_c + p_s = \int_y^{h_2} (\sigma - \rho) C g \cos\theta \, dy \,. \tag{5.41}$$

In eqn (5.40) the first member includes in their respective order the collisional stress, kinetic stress caused by the passage of one particle from one layer to the others, static stress due to friction in contacts and that due to turbulent mixing; the first member of eqn (5.41) includes in their respective order static and dynamic pressure. Moreover, σ is the particle density, ρ is the density of the water, θ is the bed slope, C is the solid concentration, g is gravitational acceleration and p' is the excess of fluid pressure compared to the static value, which can be expressed:

$$p' = g\cos\theta(\sigma - \rho) \int_y^{h_2} C dy \,. \tag{5.42}$$

Solid concentration at the bed, indicated by C^*, corresponds to a configuration in which the particles are highly cohesive, therefore, on the bed, velocity must be null because flow cannot exist at such high concentrations and the same applies to the dynamic pressure p_c, therefore, the entire excess of pressure created in this area is caused by static pressure p_s alone. However, in order to transmit static pressure, the particles must be continuously in contact with one another, despite the fact that their relative position changes, and as, in addition, solid concentration by volume diminishes towards the top, at a certain height, indicated for convenience's sake by y_l, to which a particular solid concentration by volume corresponds (indicated by C_l) according to the concentration distribution in the y direction, this condition will no longer occur, thus annulling static pressure. The value of C_l depends on the grain-size distribution of the solid particles in the y direction and is generally equal to 0.5 [41]. The distribution of p_s is not yet known and different formulations have been proposed by different authors; in order to respect the previous characteristics, the unified inertial model assumes the following function [33, 69]:

$$p_s = \begin{cases} 0 & \text{for } C \le C_l \\ \dfrac{C - C_l}{C^* - C_l} p' & \text{for } C > C_l \end{cases} \,. \tag{5.43}$$

For static stress, a Coulombic formula is assumed, where ϕ is the material's angle of internal friction:

$$\tau_s = p_s \tan\phi \,. \tag{5.44}$$

For turbulent stress, the theory of turbulent mixing is assumed:

$$\tau_t = \rho_m l^2 \left(\frac{du}{dy}\right)^2, \tag{5.45}$$

where $\rho_m = [(\sigma - \rho) + \rho]$ is the mixture's apparent density; because in this layer the mixing length l would be of the order of the voids between particles, one assumes:

$$l = d\left[\left(\frac{C^*}{C}\right)^{1/3} - 1\right]. \tag{5.46}$$

For collisional and kinetic stress, Takahashi et al. [69], for a bidimensional and dry flow of granular materials, propose:

$$\tau_c = \frac{4}{5} C^2 g_0 \sigma d_p (1+e) \sqrt{\frac{T}{\pi}} \left(\frac{du}{dy}\right) \tag{5.47}$$

$$\tau_k = \frac{\sigma d_p}{3 g_0} \sqrt{\frac{T}{\pi}} \left(\frac{du}{dy}\right), \tag{5.48}$$

with d_p particle diameter and e the restitution coefficient, ratio between the difference in velocity of two particles after a collision and the difference between their initial velocities (e=0 in the case of inelastic collision and e=1 for elastic collisions); T is granular temperature and g_0 is a function of the solid concentration by volume and they have the following expressions [69]:

$$g_0 = 1 + \frac{1}{\left[\left(C^*/C\right)^{1/3} - 1\right]} \tag{5.49}$$

$$T = \frac{d_p^{\,2}}{15(1-e)} \left(\frac{du}{dy}\right)^2. \tag{5.50}$$

In the upper layer the stress balance is as follows:

$$\tau_c + \tau_t = \int_y^{h_t} \rho_m g \sin\theta \, dy. \tag{5.51}$$

For the mixing length, not having other information, it is assumed that l=ky, with k von Karman's constant.

Velocity and solid concentration distribution in the region where $C \geq C_l$, the bottom part of the lower layer, can be obtained as follows:

$$\frac{dC}{dY} = \frac{\left(sB^2C/\tan\theta\right) - B}{A\left(1 - Y + s\overline{C}\right)} \tag{5.52}$$

$$\left(\frac{\mathrm{d}u'}{\mathrm{d}Y}\right)^2 = \frac{15\sin\theta}{2C^2 g_0 B}\frac{\rho}{\sigma}\left(\frac{h}{d_p}\right)^2 \frac{1-e}{1+e}[1-f(C)](1-Y+s\overline{C}), \tag{5.53}$$

in which $Y = y/h$ and $u' = u/\sqrt{gh}$, $s = (\sigma-\rho)/\rho$, $\overline{C} = \int_{h_1/h}^{1} C\mathrm{d}Y$,

$f(C) = \dfrac{C-C_1}{C^* - C_1}$ (for the definition of constants A and B see Takahashi, [41]). In the region where $C<C_1$, the top part of lower layer, it is hypothesised that $f(C)=0$, which could cause an underestimation of the solid discharge that can increase as the slope angle decreases and the larger layer dominated by turbulent stress [70]; however, in this layer one obtains:

$$\frac{\mathrm{d}C}{\mathrm{d}Y} = \frac{\left(sB'^2 C/\tan\theta\right) - B'}{A'\left(1-Y+s\overline{C}\right)} \tag{5.54}$$

$$\left(\frac{\mathrm{d}u'}{\mathrm{d}Y}\right)^2 = \frac{15\sin\theta}{2C^2 g_0 B'}\frac{\rho}{\sigma}\left(\frac{h}{d_p}\right)^2 \frac{1-e}{1+e}[1-f(C)](1-Y+s\overline{C}), \tag{5.55}$$

(for the definition of the constants A' and B' see Takahashi, [41]).

In order to obtain the velocity and solid-concentration distribution in the upper layer, one could observe that, as the mechanisms with which turbulence originates in solid–liquid flows are considerably different from those observed for hydrodynamic streamflows (for which one generally talks of wall turbulence), it could seem incorrect to consider the mixing length from the channel bed, however, one can take an expression such as $l = k(y-\xi)$ is assumed, where k is von Karman's constant and ξ is a vertical coordinate that originates in ξ_0, which represents the distance from the bed in which a fictitious wall is posed and in relation to which the mixing length is evaluated (Figure 5.13). However, this parameter is only fictitious and for its evaluation reference is made to simple hypotheses of physical continuity, such as the continuity of the shear rate in the vertical direction in the passage to the layer in which $C=C_0$ (C_0 is the solid concentration in volume corresponding to the height ξ_0) to that above, in which it is assumed that $f(C) = \dfrac{C-C_1}{C^* - C_1} = 0$ [41]. As the particles are in turbulent suspension, in the same way as the case of suspended-load transport one has:

$$\frac{C}{C_0} = \left[\frac{\xi_0}{y+\xi_0-(h-h_1)}\right]^{w_s/ku_1^*} \tag{5.56}$$

$$\left(\frac{\mathrm{d}u'}{\mathrm{d}Y}\right)^2 = \frac{\int_Y^1 (sC+1)\sin\theta dY}{\frac{4}{5}\frac{\sigma}{\rho}\left(\frac{d_p}{h}\right)^2 \frac{1+e}{\sqrt{15\pi(1-e)}} C^2 g_0 + (1+sC)l'^2}, \tag{5.57}$$

in which w_0 is the sedimentation velocity of the particles and $u_1^* = \sqrt{gh_1 \sin\theta}$.

5.4.8 Debris-flow hydraulics

In order to describe the behaviour of a debris flow, one must know the constitutive equation or resistance law to use for each type of flow, which would also require the estimate of the coefficients of the resistance to shear or that of flow behaviour. These coefficients have wide fields of variation and therefore in evaluating them a considerable error is committed. In particular, the flow-behaviour index η and the consistency index μ_1 of Chen's GVF model, closely depend on flow depth, the velocity and concentration of the particles [49]. It was observed that the flow-behaviour index μ_1, calculated during later laboratory simulations, ranged from 9.9 to 4415 [49], with a variation of three orders of magnitude, which indicates that the calculation of velocity made later using these models could cause considerable errors. On the other hand, some empirical equation of velocity is necessary in any debris-flow disaster-forecasting project; although the hydraulics of a debris-flow is not theoretically comparable to that of a traditional water flow, due to the high sediment concentration and the interactions between the solid particles, some experimental data [71] show that it is possible to use the resistance laws of conventional turbulent flows, in order to estimate the behaviour of such phenomena. Being more diluted than the head, the main body of the flow could be conveniently described by a traditional law of resistance such as that of Manning or Chezy. The hydraulic formations generally use momentum balances averaged on the flow depth, with the hypothesis of a thin layer or with the approximation of the kinematic wave.

Therefore, by following a hydraulic approach, the mean flow velocity V (m/s) can be simply expressed in function of the flow depth h (m), the slope S ($\sin\theta$) and a coefficient of overall resistance C_r; in a generalised relationship this dependence could be posed in the following form:

$$V = C_r h^\alpha S^\beta , \tag{5.58}$$

where the exponents depend on the approach used in the following way:

Table 5.7: Hydraulic approach, values of the coefficient and exponents of the main models.

Flow type	C_r	α	β
Newtonian laminar flow	$(1/3)\rho g/\mu$	2	1
Dilatant grain shearing	ξ	1.5	0.5
Newtonian turbulent flow: Manning	$1/n$	0.67	0.5
Newtonian turbulent flow: Chezy	C'	0.5	0.5

ρ being the mixture density (kg/m^3), μ (kg/s m) the dynamic viscosity, ξ (s m$^{-1/2}$) is a coefficient that depends on concentration and the dimension of the granules, n (s/m$^{1/3}$) and C' (m$^{1/2}$/s) are the Manning and Chezy coefficients, respectively. In agreement with experimental results [71] one obtains that flow resistance increases with the solid concentration of the debris-flow wave; this suggests the presence of a continuous transition in the trend of the flow resistance, passing from water flows to fluvial sediment-transport processes on steep flows, up to debris-flows.

One limit of the approaches that use this hydraulic formulation is that they use empirical coefficients in order to express in a parametric form the distribution of the momentum and energy dissipation in a fully developed motion regime, without, however, considering triggering dynamics, deformation, and deposition [3].

5.5 Conclusions

A critical analysis has been made of the various models currently available for describing the mechanisms of debris-flow mechanics in uniform motion. With regards to the study of the models that characterise the mechanics of a debris-flow in uniform motion, by following the classification of the regimes typical of debris-flows proposed by Jan and Shen [49], velocity profiles were obtained for each regime for a steady and bidimensional flow, with the hypothesis of a uniform mixture and constancy of the relative parameters and coefficients. It has also been highlighted how with the variation in flow characteristics that occur in various areas, authors in the literature have proposed different and apparently conflicting models: this is fully justifiable if one considers the difficulty in extending a single model to the complex and varied spectrum of events that occur in nature, thus avoiding excessively generalised formulations. Moreover, the rheological behaviour of a mixture depends on the hydrodynamic characteristics of the flow and the characteristics of the sediment, which can vary from one flow to another; lastly, it should be pointed out that many models are only applicable to the particular cases in which they have been obtained and they therefore intrinsically lack generality. In certain cases a one–layer model is unable to adequately describe the entire thickness of the flow, and therefore it has recently become common to work with multi layer models that combine two or more constitutive relationship in order to find an adaptation to experimental velocity profiles. The coefficients, such as the flow-behaviour index or relative parameters, depend strongly on many factors such as the dimension, concentration and distribution of the sediments, as well as the flow regime (laminar or turbulent). The determination of this dependency constitutes one of the aims of current studies [49]. If the debris-flow is then treated as a nonuniform mixture, in order to resolve the problem a diffusion equation is required, in addition to that of mass conservation and momentum; unfortunately very little information is available on the diffusion mechanism and deposition (or erosion) processes. The current of hydraulic models, whose local empirical equations of mean velocity are useful and directly applicable in debris-flow disaster-prevention models, is equally significant.

Chapter 6

Debris-flow modelling: a general outline of the motion problem

6.1 Introduction

Although examining past events is important in forecasting potential future debris flows, theoretical modelling remains essential to understanding the phenomenon and therefore selecting the most suitable mitigation measures to adopt.

The analysis of the various models proposed (see Chapter 5) and used in the literature highlights two matters: firstly, that the great variability of the phenomenon necessarily requires the use of a number of different rheological models and secondly that nevertheless, studies on debris-flow dynamics often do not dedicate adequate attention to all the possible effects of the stresses acting simultaneously. Basins characterised by compact silty or clayey soils are more likely to produce flows with mainly viscous or viscoplastic behaviour and can therefore be described by Herschel-Bulkley-type models. The presence of small percentages of cohesive solid materials would suggest that mudflows are less probable in mountainous areas, given also the threshold concentration value indicated by Coussot (ratio between clay concentration and granules concentration equal to 0.10). Mountain basins, in fact, have grain sizes of the order of 10^{-1} – 10 m, and it is therefore very probable that plastic and dilatant-type constitutive relationships can better represent the flows that occur in these conditions. In fact, in mountain areas the material involved in transport processes on steep slopes is usually coarse and scarcely cohesive, and the flow may reach high dynamicity, which would suggest in the rheological studies the use of dilatant-type models for the characterisation of solid–liquid mixtures. However, the drawback of the dilatant model as originally proposed by Bagnold is the existence of certain empiricisms, which make it inadequate for describing the behaviour of the entire mixture alone, the introduction of terms independent of the shear rate being necessary. It should also be pointed out that in order to obtain an instrument capable of adequately estimating the behaviour of granular mixtures for various concentrations, turbulent stresses must also be included in formulating the model. The key problem in the study of solid–liquid mixture mechanics lies in identifying the constitutive relationships that describe the fluxes of mass, momentum and energy and understanding the mechanisms that govern the development of the stresses.

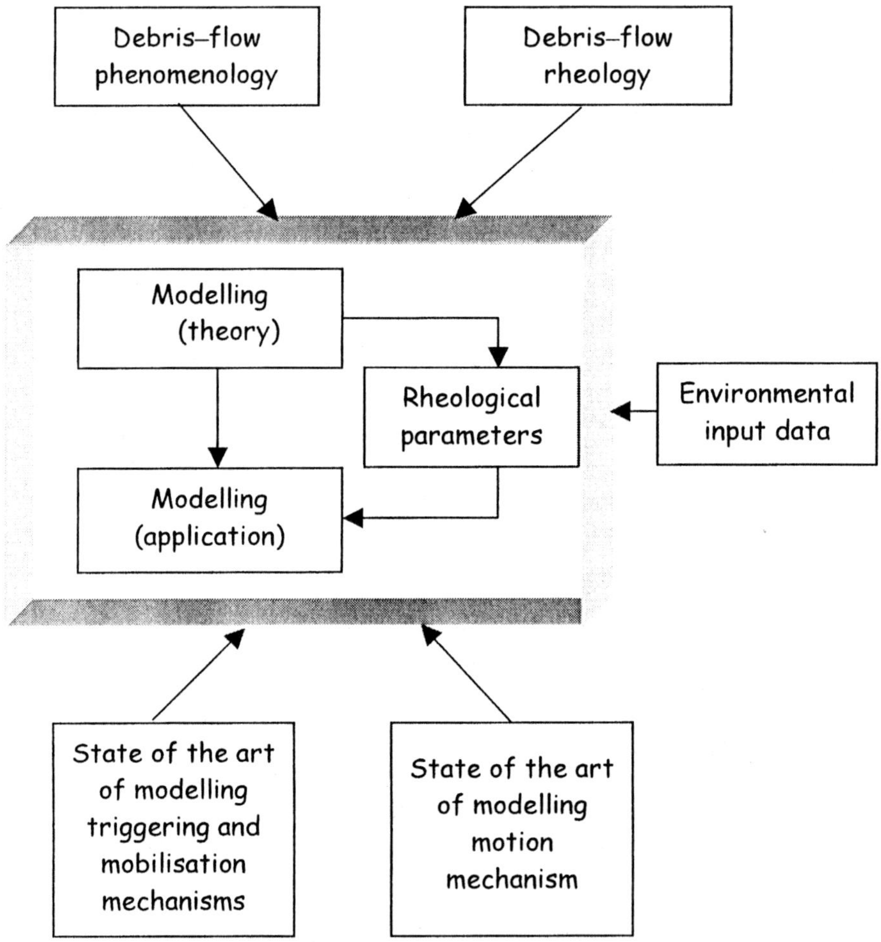

Figure 6.1: In the chart, showing a breakdown of the study approach adopted in this book, the grey frame highlights the analysis field of debris-flow modelling, the topic that will be dealt with in this chapter.

In this chapter, the problem of debris-flow modelling is tackled from a general point of view, starting from a study of the system's mass and momentum balances and the analysis of the different types of stress acting within the flow.

6.2 Debris-flow as fluid-mechanical phenomenon

A solid–liquid mixture can be described as a collection of discrete solid particles dispersed within a fluid in such a way that, being very frequent, the direct interactions between the single granules determine the importance of interparticle forces, friction and collisions [17]. Solid–liquid mixtures pertain to the vast class of dispersed two-phase systems constituted by a solid and a liquid component and include the extremes cases of granular mixtures, in which the solid phase entirely influences the behaviour of the entire system on the one hand, and dilute suspensions, in which the opposite occurs and the fluid phase plays a dominant role, on the other. However, debris flows are solid–liquid mixtures in which both the solid and the liquid components play a decisive role in characterising the flow. In any case, the overall behaviour of these systems is very complex and its understanding requires a synergy of interdisciplinary contributions from traditional fluid mechanics, as well as the theory of plasticity, soil mechanics, rheology, and the kinetic theory of gases [72]. With the establishment of microscopic kinetic models (see Section 3.5.1) following the fundamental contribution made by Ogawa [43], it became common to study the dynamic behaviour of solid–liquid mixtures from a continuous point of view, as generally happens in fluid mechanics, considering the single grains as the molecules of a granular "fluid". For this reason, Campbell [35] defined the flow of solid–liquid mixtures as a *fluid-mechanic phenomenon.* It must, however, be pointed out that from a mechanical point of view there are a number of substantial differences between two-phase mixtures and simple fluid systems [35]:

- a grain of sand, taken as an archetype of a granular particle, is approximately 10^{18} times more voluminous and heavy than a molecule of water; this difference has an important consequence on the applicability of the continuity hypothesis. Due to the size of the granules, the number of molecules per unit volume of a granular system is much smaller than in a fluid system; one millimetre cube of water contains approximately 10^{19} molecules, whereas the same volume of sand could contain as little as 10 granules. In order to be able to use the hypothesis of a continuous medium, we must define the precise values of each property, in terms of measurable physical quantities on a scale larger than a molecular one and that varies in a uniform and continuous way throughout the system;
- unlike molecules, the granules undergo inelastic collisions that are assumed to be binary, thus causing a progressive loss of energy in the form of heat released during the collision;

- moreover, we cannot consider that the particles have the same shape, in fact, within a granular system we must expect to find a wide spectrum of granule sizes, which introduces significant complications to their theoretical description;
- because the grains are not spherical and have a rough surface, friction shear forces exist that mean that the granule–granule interaction is not central, thus causing the existence of a torque that causes rotation on the single particles;
- last, but not least, the existence of an interstitial fluid considerably alters collisional dynamics.

6.3 General outline of the problem

The key problem of solid–liquid mixture mechanical studies lies in identifying the constitutive relationship that describes the fluxes of mass, momentum and energy and the understanding of the mechanisms that govern stress development. There are two main schemes for representing motion in debris-flow studies: one assumes the hypothesis of debris-flows as a homogeneous mixture and the other keeps the solid and liquid phases distinct and assumes a two-phase mixture. The equations for the mass and momentum balances in each approach and their respective strong and weak points are discussed below.

6.3.1 Mass and momentum balances for homogeneous mixtures

Motion equations for a homogeneous water–sediment flow are initially obtained for a mixture in motion along a level surface by means of mass and momentum balances for a control volume coinciding with an infinitesimal fluid element [73]. For simplicity, it is assumed that the field of motion is bidimensional and that all dominions, including control volumes, have a unitary and constant thickness in the y direction orthogonal to the ideal surface, which represents the channel bed. A steady flow is assumed, for which velocity and pressure at each point do not vary in time, whereas, in general, there will be a spatial variation in these quantities through the flow field. The no-slip hypothesis is also introduced by which the mixture in direct contact with the solid boundaries of the channel is at rest and therefore the velocity component tangent to the surface is null. All mixture properties are assumed uniform and constant throughout the flow field. In particular, it is assumed that the process is isothermal, adiabatic and takes place at a constant temperature (in other words heat transfer is not involved) and that the mixture is incompressible and therefore density variations with pressure are negligible. In such conditions, the continuity equation expresses the mass conservation principle, which, for an infinitesimal control volume in a steady regime, means that inflow and outflow mass in the time unit are equal (Figure 6.2):

$$\rho u \mathrm{d}z + \rho v \mathrm{d}x = \left[\rho u + \frac{\partial(\rho u)}{\partial x}\mathrm{d}x \right]\mathrm{d}z + \left[\rho v + \frac{\partial(\rho v)}{\partial z}\mathrm{d}z \right]\mathrm{d}x\,, \tag{6.1}$$

where ρ is mixture density and u and v are the velocity components in a longitudinal x and orthogonal z direction in relation to the channel bed, respectively. By dividing by *dx dz* we obtain:

$$\rho\left(\frac{\partial u}{\partial x} + \frac{\partial v}{\partial z}\right) = 0\,. \tag{6.2}$$

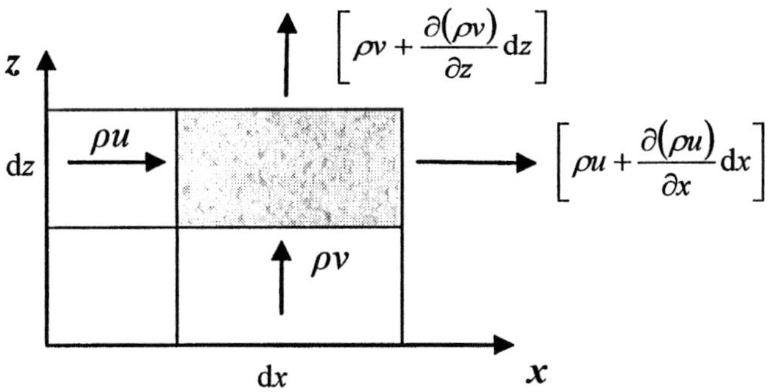

Figure 6.2: Mass balance for an infinitesimal control volume.

The equation that governs the distribution of the velocity component u, directed along the x axis, can be calculated on the basis of Newton's second law, interpreted as the momentum conservation principle: in steady flows, the variation in momentum fluxes in and out from the control volume is equal to the resultant of the forces acting in the same direction and on the same volume [73].

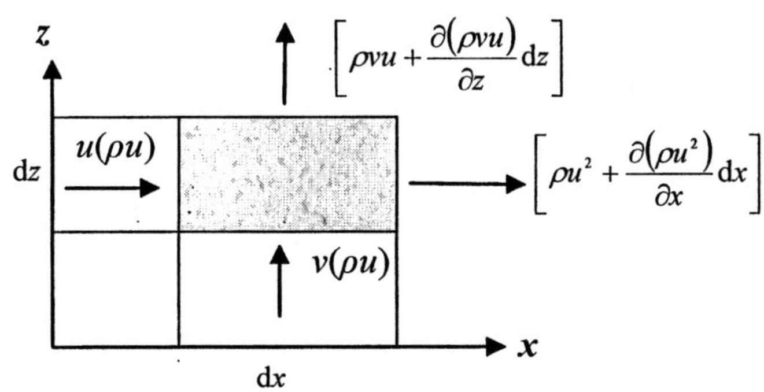

Figure 6.3: Momentum balance for an infinitesimal control volume.

Therefore, the momentum balance in the x direction gives (Figure 6.3):

$$\Sigma F_{\mathrm{x}} = \left\{\left[\rho u^2 + \frac{\partial(\rho u^2)}{\partial x}\mathrm{d}x\right]\mathrm{d}z - \rho u^2 \mathrm{d}z\right\} + \left\{\left[\rho v u + \frac{\partial(\rho v u)}{\partial z}\mathrm{d}z\right]\mathrm{d}x - \rho v u \mathrm{d}x\right\}. \quad (6.3)$$

For an incompressible flow, and by exploiting the continuity equation, eqn (6.2), we can obtain:

$$\left(\rho u \frac{\partial u}{\partial x} + \rho v \frac{\partial u}{\partial z}\right)\mathrm{d}x\mathrm{d}z = \Sigma F_{\mathrm{x}} \quad (6.4)$$

The resultant of the forces in the x direction, F_{x} in the first member of the equation above, includes both body forces (due to the gravitational field) and superficial ones (pressures and stresses) as shown in Figures 6.4 and 6.5.

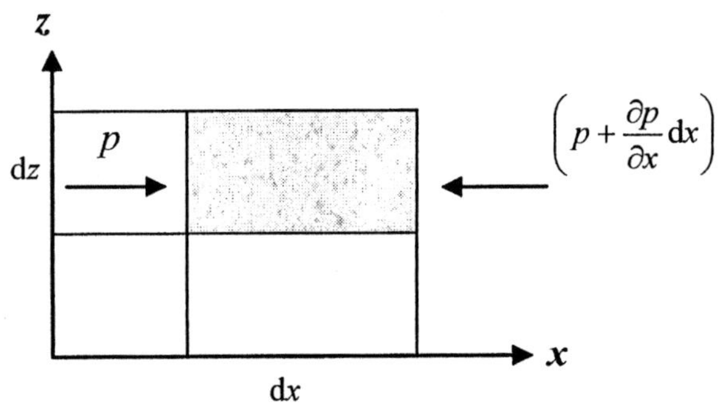

Figure 6.4: Pressure forces acting in the x direction on an infinitesimal control volume.

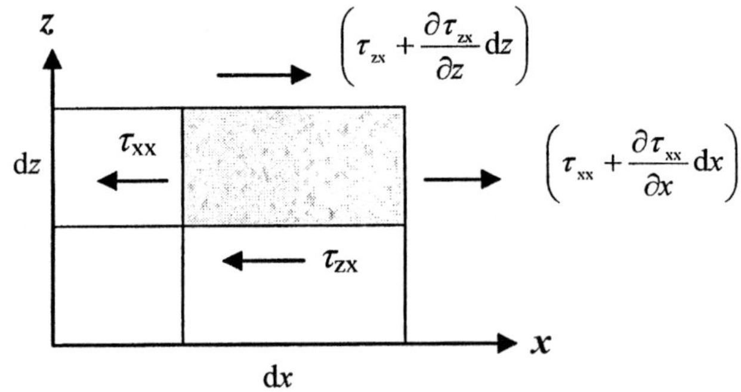

Figure 6.5: Components of the stress tensor acting in the x direction on an infinitesimal control volume.

By indicating with τ_{zx} the stress exerted in the x direction on a surface orthogonal to the z axis, with τ_{xx} the stress exerted in the x direction on a surface orthogonal to the x axis and with p the fluid pressure, the force balance gives:

$$\Sigma F_x = p\mathrm{d}z - \left(p + \frac{\partial p}{\partial x}\mathrm{d}x\right)\mathrm{d}z + \left(\tau_{zx} + \frac{\partial \tau_{zx}}{\partial z}\mathrm{d}z\right)\mathrm{d}x - \tau_{zx} + \left(\tau_{xx} + \frac{\partial \tau_{xx}}{\partial x}\mathrm{d}x\right)\mathrm{d}z - \tau_{xx} \cdot \tag{6.5}$$

By dividing by $dx\ dz$, the momentum balance in the x direction gives:

$$\rho\left(u\frac{\partial u}{\partial x} + v\frac{\partial u}{\partial z}\right) = -\frac{\partial p}{\partial x} + \left(\frac{\partial \tau_{zx}}{\partial z} + v\frac{\partial \tau_{xx}}{\partial x}\right). \tag{6.6}$$

In the same way, by expressing the variation of the momentum flux in the z direction, one obtains an expression similar to that above:

$$\rho\left(u\frac{\partial v}{\partial x} + v\frac{\partial v}{\partial z}\right) = -\rho g - \frac{\partial p}{\partial z} + \left(\frac{\partial \tau_{zx}}{\partial x} + \frac{\partial \tau_{zz}}{\partial z}\right), \tag{6.7}$$

in which τ_{xz} is the stress exerted in the z direction on a surface orthogonal to the x axis, τ_{zz} the stress exerted in the z direction on a surface orthogonal to the z axis, p is the fluid pressure and g is the gravitational acceleration.

In short, the vectorial equations that express mass and momentum conservation for a homogeneous water–sediment flow are the following:

$$\nabla(\rho \cdot \mathbf{u}) = 0 \tag{6.8}$$

$$\rho \cdot \mathbf{u} \cdot \nabla \mathbf{u} = \rho \mathbf{g} - \nabla p + \nabla \mathbf{T} = 0\,, \tag{6.9}$$

in which ρ is the density of the fluid–sediment mixture, which is assumed to be homogeneous, $\mathbf{T}$ is the stress tensor, p is the pressure, g is the gravitational acceleration, $\mathbf{u}$ is the velocity vector (vectorial sum of component u in the x direction and component v in the z direction).

6.3.2 Mass and momentum balances for two-phase mixture

The mass and momentum balance equations for a debris flow can be borrowed, by making suitable alterations, from the continuous theory of two-phase mixtures, initially developed by Goodman and Cowin [51], which represents a relatively mature field of research. It is assumed that the material properties of the mixture are not constant, but can be represented by continuous functions of the medium's punctual state; therefore the kinematic variables $v(x, t)$ and $\rho(x, t)$, which represent the volume and mass distribution functions, respectively, defined in region V occupied by the mixture, are introduced. The relative volume V_s and mass M_s of the solid component are therefore given by:

$$V_s = \int_V v(\mathbf{x}, t)\mathrm{d}V \tag{6.10}$$

$$M_s = \int_V \rho_s v \mathrm{d}V\,. \tag{6.11}$$

The fundamental question of representing the motion of two–phase mixtures can also be integrated using a micromechanical approach, which therefore aims to describe the macroscopic behaviour of the mixture, starting from an analysis of the interaction of the phases at a single granule level. The liquid can therefore be considered as a continuum, whose motion is described by the local values of the various quantities, whereas the solid phase is assimilated to a discrete system of rigid particles, for each one of which the motion induced by gravity and by the interactions with the fluid phase are studied. Given the considerable complexity of the mathematical problem arising from this model, the micromechanical approach, developed by Batchelor in 1970, resorts to a description based on mean quantities [72]. This formulation is based on the ergodic hypothesis and the spatial-homogeneity hypothesis; with the former, the overall mean of a certain quantity and a spatial mean performed on a coordinate in relation to which the mixture is statistically steady are assumed to be equal. For the second hypothesis, we must consider a length l in relation to the mean intergranular distance, but such that on l the statistical properties of the mixture, posed as overall means, do not vary considerably [72]. The balance equations for each phase are obtained by averaging them with suitable balances performed on a control volume of spatial scale l.

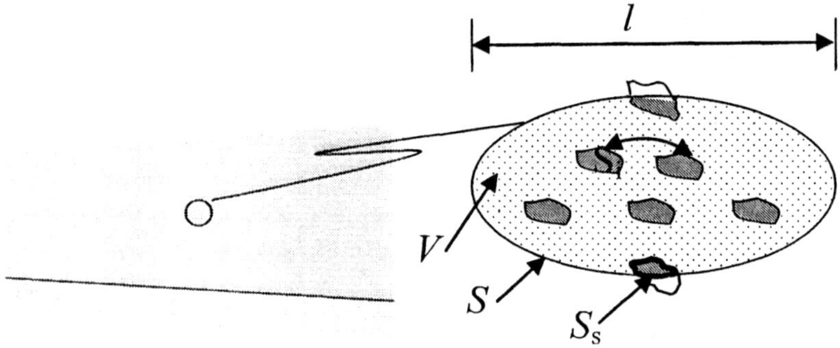

Figure 6.6: Definition diagram for the mixture balance in the micromechanical approach [72].

We then consider a fixed volume V of spatial scale l and we use V_k to indicate the portion occupied by the generic phase k (solid or liquid), whose boundary will be generally constituted by S_k, which is that part of the boundary of V belonging to phase k, and by S_i relative to the interfaces contained in the chosen volume. Jenkins and Savage [74] derived a general balance equation for any conserving quantity ψ referring to the unit of mass in the generic phase k and relative to the entire volume V:

$$\frac{\partial}{\partial t}\int_{V_k}(\psi^k\rho_k)dV_k = \int_{S_k}(-\psi^k\rho_k\mathbf{v}^k+\mathbf{\Lambda}^k)\cdot\mathbf{n}^k dS_k + \int_{S_i}[-\psi^k\rho_k(\mathbf{v}^k-\mathbf{v}^i)+\mathbf{\Lambda}^k]\cdot\mathbf{n}^k dS_i + \int_{V_k}(\Theta^k\rho_k)dV_k \quad (6.12)$$

The equation above contains the local and instantaneous values of the quantities referring to the generic phase k (ρ is density and $\mathbf{v}$ velocity), and $\mathbf{\Lambda}$ represents the nonconvective flux of ψ, Θ is the specific value of the velocity of production of ψ within V, and $\mathbf{n}$ is the normal unitary vector assumed to be positive if directed outwards in relation to the generic phase. The balance of ψ^k in V_k depends on the three contributions to the second member that represent, in order:

- the total transport of ψ^k through the fixed boundary S_k due to both the convective effect and any nonconvective effects (for example, the stresses acting on the boundary S_k in the case in which ψ^k represents momentum);
- the transport of ψ^k through the interfaces within V that are in motion with velocity $\mathbf{v}^i$;
- the velocity of production of ψ^k inside V due to the effect of external fields, collisions and contact actions.

We assume $\nu_k = V_k/V$, which indicates the volumetric fraction of phase k; moreover, the instantaneous value of the spatial mean of any quantity f^k is expressed by:

$$\langle f^k\rangle = \frac{1}{V_k}\int_{V_k} f^k dV_k . \quad (6.13)$$

The form averaged on the volume V_k of eqn (6.12), and then divided by V, is the following [72]:

$$\frac{\partial}{\partial t}\left(\nu_k\langle\psi^k\rho_k\rangle\right)+\nabla\cdot\left(\nu_k\langle\psi^k\rho_k\mathbf{v}^k\rangle\right)=\nabla\cdot\left(\nu_k\langle\mathbf{\Lambda}^k\rangle\right)+\nu_k\langle\Theta^k\rho_k\rangle + \frac{1}{V}\int_{S_i}[-\psi^k\rho_k(\mathbf{v}^k-\mathbf{v}^i)+\mathbf{\Lambda}^k]\cdot\mathbf{n}^k dS_i \quad (6.14)$$

- When ψ^k=1 the quantity that is conserved is the mass and there is neither a diffusive flux of mass through the boundaries, nor any mass production mechanism within the volume and the velocity of the interfaces equals the velocity of phase k.
- However, when ψ^k= v^k we obtain the momentum balance and the non convective flux is the stress tensor $\mathbf{T}$, whilst the momentum production term is associated with the action of the field of external forces and intergranular interactions.

If we apply this method for determining the mass and momentum balances of a debris flow, it is possible to deduce separate yet closely linked equations for each mixture constituent. The angular momentum equation could also be formulated in the same way, however, by assuming the stress tensors are symmetrical, this is not necessary [3]. From the continuous theory, whose basic hypotheses were illustrated previously, we obtain the momentum and mass balance equations for the solid and liquid constituents:

$$\frac{\partial(\rho_s \nu)}{\partial t} + \nabla \cdot (\rho_s \nu \mathbf{u}^s) = 0 \tag{6.15}$$

$$\frac{\partial[\rho_f(1-\nu)]}{\partial t} + \nabla \cdot [\rho_f(1-\nu)\mathbf{u}^f] = 0 \tag{6.16}$$

$$\rho_s \nu \left(\frac{\partial \mathbf{u}^s}{\partial t} + \mathbf{u}^s \cdot \nabla \mathbf{u}^s \right) = \rho_s \nu \mathbf{g} - \nu \nabla P_f - \mathbf{F}_i + \nabla \mathbf{T}^s \tag{6.17}$$

$$\rho_f(1-\nu)\left(\frac{\partial \mathbf{u}^f}{\partial t} + \mathbf{u}^f \cdot \nabla \mathbf{u}^f \right) = \rho_f(1-\nu)\mathbf{g} - (1-\nu)\nabla P_f + \mathbf{F}_i + \nabla \mathbf{T}^f . \tag{6.18}$$

$\mathbf{u}^k$ indicates the mean velocity of the generic phase, so that $\mathbf{u}^k = \langle \mathbf{v}^k \rangle$, $\mathbf{g}$ is gravitational acceleration, $\mathbf{T}^s$ and $\mathbf{T}^f$ are the stress tensors for the solid and fluid phases, $\mathbf{F}_i$ is the interaction force per unit volume that results from the exchange of momentum between the solid and liquid constituents and P_f is the spatial mean of the instantaneous pressure acting in the fluid. The equations above are posed in relation to one another by the condition on the volumetric fractions $\nu_s + \nu_f = 1$; by posing $\nu_s = \nu$ it follows that $\nu_f = (1-\nu)$. The sum of eqns (6.15) and (6.16) and of eqns (6.17) and (6.18) lead to the mixture's mass and momentum conservation equations, respectively:

$$\frac{\partial \rho}{\partial t} + \nabla \cdot (\rho \mathbf{u}) = 0 \tag{6.19}$$

$$\rho \left(\frac{\partial \mathbf{u}}{\partial t} + \mathbf{u} \cdot \nabla \mathbf{u} \right) = \rho \mathbf{g} - \nabla P_f + \nabla \cdot (\mathbf{T}^s + \mathbf{T}^f + \mathbf{T}'), \tag{6.20}$$

with

$$\mathbf{T}' = -\rho_s \nu (\mathbf{u}^s - \mathbf{u}) \cdot (\mathbf{u}^s - \mathbf{u}) - \rho_f (1-\nu)(\mathbf{u}^f - \mathbf{u}) \cdot (\mathbf{u}^f - \mathbf{u}),$$

where $\rho = \rho_s \nu + \rho_f(1-\nu)$ and $\mathbf{u} = [\rho_s \nu \mathbf{u}^s + \rho_f(1-\nu)\mathbf{u}^f]/\rho$, the density and velocity of the centre-of-mass of an element of the mixture's volume, respectively.

If we now compare eqns (6.19) and (6.20), in the steady case, with the vectorial equations, eqns (6.8) and (6.9), obtained for the motion field of a homogeneous water–sediment flow, we can observe how, by substituting $\mathbf{T}^s+\mathbf{T}^f$ by $\mathbf{T}$ and P_f by p, eqns(6.19) and (6.20) effectively reflect those equations, if we exclude $\mathbf{T}'$, which represents the contribution linked to the relative motion of the two constituents in relation to the mixture as a whole. From a mathematical point of view, this term is explained by observing that the convective terms of the first member of eqns (6.17) and (6.18) added together do not simply give $(\mathbf{u}\cdot\nabla\mathbf{u})$, the convective acceleration of the mixture. From a physical point of view, $\mathbf{T}'$ indicates that the stresses in a two-phase debris flow, modelled as a homogeneous system, are more complex than those obtained simply by adding $\mathbf{T}^s$ and $\mathbf{T}^f$. Two-phase models have certain advantages over those that assume a homogeneous system: the former, in fact, not only include distinct tensors in order to represent the stresses in the solid and fluid phase, but also an interaction force between the two phases, thus clearly describing its different physical nature, instead, in the latter, the effect of the solid and liquid phases and their interactions are amalgamated in a single contribution and thus often require a number of corrective parameters [3]. Given that solid−fluid interactions vary from one point of the debris-flow to another and play an important physical role, their explicit representation is certainly desirable [38] and it seems to require further research; that being so, the organization of the previously mentioned physical mechanisms in this chapter will follow, nevertheless, an intermediate approach.

6.4 Balance equations in uniform motion

At this point, we can consider a uniform and bidimensional motion that takes place asymptotically uphill behind the head of a debris flow with a depth h, that proceeds at a constant velocity U, with $\mathbf{u}^s=(u^s, v^s)$ and $\mathbf{u}^f=(u^f, v^f)$ components of velocity of the solid and liquid phase, respectively, directed along x and z (Figure 6.7).

As motion is steady and uniform, the continuity equations require $\nu=\nu(z)$, $u^s=u^s(z)$ and $u^f=u^f(z)$, therefore the velocity profiles and volumetric fraction only vary in a direction orthogonal to the surface, in other words, they are a functions of z:

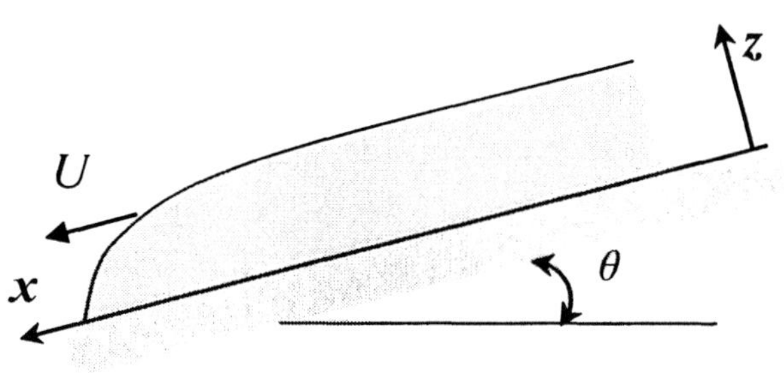

Figure 6.7: Definition diagram for the balance equations in uniform flow.

$$\frac{\partial(u_s\nu)}{\partial x}=\frac{\partial[(1-\nu)u_f]}{\partial x}=0\,. \tag{6.21}$$

It therefore follows that the solid and liquid discharges are constant, that is:

$$q_s = \int_0^h \nu u_s dz = \cos t$$
$$q_f = \int_0^h (1-\nu) u_f dz = \cos t. \tag{6.22}$$

Often, in the technical literature on debris flows (Seminara and Tubino, [72]; Takahashi, [21]) it is arbitrarily assumed that on average there is no relative motion of the two phases, that is $u_s = u_f$, it therefore follows that:

$$\rho_s \nu u_s + \rho_f (1-\nu) u_f = [\rho_s \nu + \rho_f (1-\nu)] u = \rho u, \tag{6.23}$$

and that $u_s = u_f = u$. Equation (6.20) for the uniform motion of debris flow can therefore be written thus:

$$0 = \rho \mathbf{g} - \nabla P_f + \nabla \cdot (\mathbf{T}^s + \mathbf{T}^f). \tag{6.24}$$

In the previous equation we have considered that the inertia associated with the mean component of the motion of both the solid and liquid phases is null, and that $\mathbf{T}'=0$ having assumed that on average there is no relative motion of the two phases; however, in order not to neglect the interactions between the solid and liquid phase, eqn (6.24) must be accompanied by the balance equation for the fluid phase (6.18):

$$\nabla P_f = \rho_f \mathbf{g} - (1-\nu)^{-1} (\mathbf{F}_i + \nabla \mathbf{T}^f). \tag{6.25}$$

Lastly, by projecting eqns (6.24) and (6.25) along x and z, we obtain the balance equations in the form suitable for uniform motion (Figure 6.7):

$$\frac{\partial \tau_{zz}}{\partial z} = \frac{\partial P_f}{\partial z} + [(\rho_s - \rho_f)\nu + \rho_f] g \cos\theta \tag{6.26}$$

$$\frac{\partial \tau_{zx}}{\partial z} = -[(\rho_s - \rho_f)\nu + \rho_f] g \sin\theta \tag{6.27}$$

$$\frac{\partial P_f}{\partial z} = -\rho_f g \cos\theta + (1-\nu)^{-1} \left(\frac{\partial \tau_{zz}^f}{\partial z} + F_{iz} \right) \tag{6.28}$$

$$0 = \rho_f g \sin\theta + (1-\nu)^{-1} \left(\frac{\partial \tau_{zx}^f}{\partial z} + F_{ix} \right). \tag{6.29}$$

In the engineering literature, (Takahashi, [21, 33, 63]; Egashira et al., [75]; Julien and O'Brien, [32]) from the point of view of the contributions to mixture dynamics, the equations above are treated as follows:

- Equation (6.28), which represents the equation for the fluid phase along z, is simplified by observing that in a steady and uniform flow $F_i=0$ [21] and by explicitly ignoring the contribution $\partial \tau_{zz}^f / \partial z$, associated with the turbulence of the fluid and to the alteration that it produces on the collisional dynamics of the particles; however, these effects must be implicitly considered in the formulation of a suitable rheological equation for the combined tensor of the solid–fluid stresses ($\partial\tau_{zz}/\partial z$). In short, with the previous assumptions, one obtains:

$$\frac{\partial P_f}{\partial z} = -\rho_f g \cos\theta, \tag{6.30}$$

which represents the characteristic hydrostatic distribution of the mean pressure in the fluid phase, when the other contributions connected with it are neglected, to a first approximation.

- Equation (6.29), which represents the equation for the fluid phase along x, is generally ignored (Takahashi, [21, 33, 63]; Egashira et al., [75]; Julien and O'Brien, [32]).

With the above-mentioned assumptions, uniform motion balance equations (6.26) and (6.27) are therefore reduced to the form:

$$\frac{\partial \tau_{zz}}{\partial z} = -\rho_f g \cos\theta + [(\rho_s - \rho_f)\nu + \rho_f]g\cos\theta = (\rho_s - \rho_f)\nu g \cos\theta \tag{6.31}$$

$$\frac{\partial \tau_{zx}}{\partial z} = -[(\rho_s - \rho_f)\nu + \rho_f]g\sin\theta, \tag{6.32}$$

which can be integrated between z and h, by assuming that the stress is null at the free surface τ_{zx} (z=h)=0 and τ_{zz} (z=h)=0:

$$\tau_{zz}(z) = -\int_z^h (\rho_s - \rho_f)\nu g\cos\theta \,\mathrm{d}z \tag{6.33}$$

$$\tau_{zx}(z) = \int_z^h [(\rho_s - \rho_f)\nu + \rho_f]g\sin\theta \,\mathrm{d}z\,. \tag{6.34}$$

At this point, the calculation of the solution for the velocity field requires the adoption of a constituent relationship for (τ_{zx}, τ_{zz}) and the integration of eqns (6.33) and (6.34) with suitable boundary conditions on the free surface and at the bed. The need to choose a constitutive relationship suitable for (τ_{zx}, τ_{zz}) requires accurate analysis of the stresses responsible for momentum transport in the mixture and relative to the solid and liquid components and their interaction; this analysis will be the subject of the next section.

6.5 Stress analysis

The wide range of debris flows that can occur in nature can be attributed to the fact that from one case to another, the dominant stresses may vary.

The application of rheological models to solid–liquid mixtures can generally be referred to two scales: on the one hand to local characteristics, which are variable in the depth and assuming that run off takes place in locally steady conditions, and on the other hand to mean flow characteristics or to the mean values of velocity and concentration in a certain section. The first approach involves the mechanical study and the understanding of local mixture behaviour and verification, using experimental data, is required for any generalisation to the entire depth and the subsequent application on a macroscopic scale [70].

Despite the fact that the mixture substantially constitutes a solid–liquid two-phase system, modelling generally excludes a true distinction of the phases, referring rather to local mean values of the velocity and concentration in volume *C*, the latter are expressed as a function of the solid and liquid volumetric fraction:

$$C(x,z) = \frac{V_{\text{solid}}(x,z)}{V_{\text{solid}}(x,z) + V_{\text{liquid}}(x,z)}. \tag{6.35}$$

By assuming that the mixture's motion takes place in homogeneous and uniform conditions, as previously mentioned, it follows that $C(x,z) \cong C(z)$ and $u(x,z) \cong u(z)$.

Figure 6.8: Stresses acting on a vertical section of a solid–liquid mixture running off in uniform motion conditions along a slope.

Referring to the bidimensional diagram in Figure 6.8, the analysis of the set of stresses acting in a water–sediment mixture subject to non-null strain, with the assumption of homogeneous system behaviour (without initially differentiating between the solid and liquid phases) and that of the validity of a superposition of effects, lead to the statement that shear stress and pressure can be expressed by the linear sum of the respective components [32, 41]:

$$\tau_{zx} = \tau = \tau_{\mu} + \tau_{c} + \tau_{s} + \tau_{co} + \tau_{t} \tag{6.36}$$

$$\tau_{zz} = p = p_{\mu} + p_{c} + p_{s} + p_{h}. \tag{6.37}$$

In the previous equations, it is therefore implicitly assumed that each stress acts independently of the presence of the others. In eqn (6.36), the following shear stresses appear in order:

- τ_μ viscous stress due to strain and the relative interaction of the fluid with sediment particles;
- τ_c dispersive stress in inelastic collisions between particles;
- τ_s Mohr–Coulomb static stress, which depends on the prolonged contacts between the various particles;
- τ_{co} stress due to cohesion (null for cohesionless materials);
- τ_t stress linked to turbulent mixing;

and by regrouping the static and cohesive terms that represent the only components independent of the shear rate, one poses:

- $\tau_y = \tau_s + \tau_{co}$ yield stress.

Similarly, in eqn (6.37), the following pressures appear in order:

- p_c dynamic pressure due to inelastic collisions;
- p_s pressure in static friction contacts between particles;
- p_h the hydrostatic pressure in the interstitial liquid;
- p_μ viscous pressure.

Generally speaking, on the one hand the stresses act simultaneously within the solid–liquid mixture, however, it could be said that the domination of one type of stress within the flow provides a good indication as to the type of flow [32], in that it conditions rheological behaviour. On the other hand, it is possible that at different depths of the same flow, different layers are characterised by different sets of stresses, which may not perfectly coincide with that chosen to represent the flow on a macroscopic level, thus making a punctual and localised analysis of the flow necessary [70]. In some cases, modelling tackles this last problem by adopting a multilayer model, which consists in ideally dividing the flow into a number of different layers, in each of which one assumes a different stress behaviour that is typical of that particular layer, as, from one layer to the next, stress could be composed of different specific contributions [33].

Although there is good agreement on how to write eqns (6.36) and (6.37) and the terms of static and cohesive stress [32, 33, 75], the way that each of the terms is expressed differs from one to another approach, thus reflecting the differences between the stress contributions considered in each model. Therefore, by referring to the numerous experimental and theoretical studies performed on the above-mentioned constitutive equations for granular flows, it is possible to identify the main functional dependencies for the various stress contributions:

$$\tau_{\mu} = \mu f_3(C)\gamma \tag{6.38}$$

$$\tau_{c} = \rho_s f_1(C) d^2 \gamma^2 \tag{6.39}$$

$$\tau_{t} = \rho l_m^{\ 2} \gamma^2 , \tag{6.40}$$

in which suitable functions of particle concentration C are used; γ is the shear rate and l_m is the turbulent mixing length. If C is greater than 0.3 the contribution associated to collisional stress becomes increasingly significant [69], and if C is less than 0.5, static stress tends to cancel itself out [7]; in an inertial and mature debris flow, concentration always ranges from 0.3 to 0.5 [41].

6.5.1 On the dispersive stress effect

Current knowledge on the mechanics of granular mixtures owes a great deal to the theory proposed by Bagnold [7], which for the first time postulated the existence of stresses that he termed *dispersive* and associated to the granular interactions. It is now commonly agreed that this type of stress constitutes one of the dominating particularities of debris flows; subsequent research has attempted to relate different concentration intervals in the flow with the different types of predominant stress in them, however, at the present time, the limit values, which have been suggested for these intervals, if indeed we can talk of limit values, are not only somewhat unclear, but also they seem somewhat arbitrary. For example, if we examine the models proposed by Takahashi [33] and Egashira et al. [75], like Franzi and Bianco [70] have recently done, it is immediately clear how different the position of the two authors is on this point (Figure 6.9). According to Takahashi [33] turbulent stress prevails when $C<0.2$ and in this existence field, which corresponds to immature debris-flows, no dispersive stresses can subsist, therefore the material is not distributed throughout the entire depth of the flow, moreover static stress cannot be neglected for $C>0.53$ (this reference value is represented by C_3), whereas for the interval in which $0.2<C<0.53$, corresponding to mature debris-flows, dispersive stress dominates.

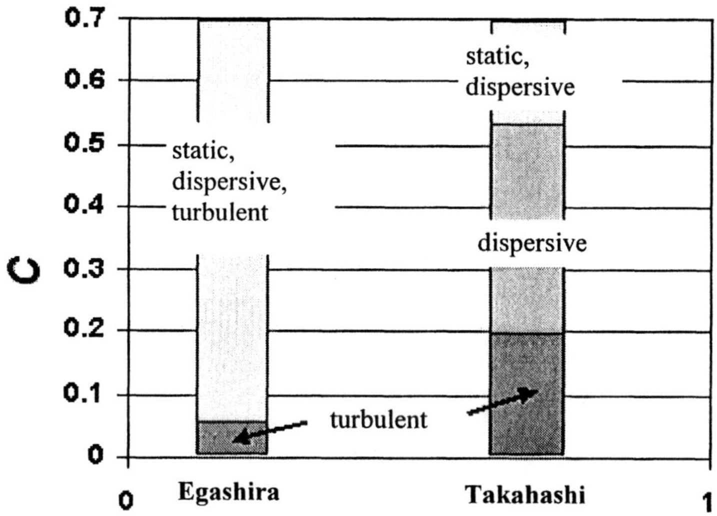

Figure 6.9: Graph showing the concentration limit values and intervals with the related characteristic stresses in the model proposed by Egashira et al. [75] and that of Takahashi [33] (see also Franzi and Bianco, [70]).

However, according to Egashira *et al.* [75], the upper C limit value of predominance of turbulent stress should be altered to 0.05, while the whole stresses could exist simultaneously for higher values of C. From the graph in Figure 6.9 one can note the role played in Egashira *et al.*'s approach by turbulent stresses even at considerable concentration values, in particular for C=0.6 there would be a substantial equivalence between turbulent and dispersive stresses. According to Chen [13] dispersive type stresses would control mixture behaviour for concentrations higher than 15%, whereas for lower concentrations the interstitial fluid would play the predominant role. In any case, these concentration thresholds cannot be considered univocal, as they depend closely on the diameter d of the grain-size used in experiments and on the height h of the solid–liquid flow. In fact, dispersive stresses also depend on the square of the diameter and, at equal shear rates, they dominate the balance of external forces if diameter is considerable, whereas they are negligible compared to turbulent stress for small diameters. To this we must add the role of flow height, in fact, at equal diameters, high h values increase the turbulence of the interstitial fluid, therefore the role of turbulent stress is significant in relation to that of dispersive stresses in suspending solid particles; this effect does not, however, influence the lower layers in which concentration is high enough to make dispersive stress only completely dominant. According to Julien [76] the inertial impact of the particles cannot be dominant when relative depth (ratio between *h* depth of the flow and *d* granule size) is greater than 25–30, for a water–sediment mixture in such conditions a turbulent flow is more likely. This is of fundamental importance because typical debris-flow heights reach 2 m, and therefore a mean particle diameter of 80 mm is necessary in order for dispersive stresses to prevail over turbulent ones. As a consequence, natural debris-flows that contain particles smaller than 80mm (gravel, sand, clay and silt) have turbulent or viscous characteristics rather than dispersive ones. With regards to this aspect, Takahashi [33] indicates that for h/d values lower than 12, there is a prevalence of dispersive-type forces, at least for concentrations higher than 0.1, whereas for higher values, there is a double-layer behaviour: a lower layer characterised by the presence of dispersive stresses alone and an upper one that involves the simultaneous presence of dispersive and turbulent stresses. Research conducted by Franzi and Bianco [70] on dispersive stresses and concerning a number of laboratory experiments, shows how more than 90% of the entire depth of flows is dominated by dispersive stress, whereas the width of the layers interested by static stresses is often lower than 10%; for small slope angle values, the width of the layer dominated by dispersive stresses is instead similar to the diameter of the particles, thus confirming the predominance of turbulent stresses in ordinary bed-load transport.

In any case, from these observations we can deduce that at high solid concentrations, the most important role is played by dispersive stresses, whereas the effect of turbulent ones is dampened by the presence of larger-sized particles. There are, however, uncertainties on the evaluation of the modifications that turbulent stresses can undergo, due to the presence of coarser particles at medium or low concentrations [70]. In short, the substantial disagreement of the formulations put forward by various authors shows that much work has yet to be done in order to give a satisfactory response to the problem of distinguishing between the various types of stresses acting in each flow layer.

6.6 Dimensionless formulation of the motion-field equations

In order to give experimental results relating to an indirect measurement, a general theoretical framework, dimensional analysis, founded on the fact that all physical laws are expressed as *dimensionally homogeneous* equations, can prove useful. The same applies for physical equations, in other words, for all those that can be directly deduced from the above-mentioned laws without the use of numerical values (*empirical constants)* having, on the contrary, their own units of measurement [77]. The term *dimensionally homogeneous* means that the change in any unit of measurement must not alter those terms of the equation in which the same quantity figures, otherwise the equation would be changed without the physical reality, that it represents, appearing changed. By dividing the value of the quantities, which describe the problem under consideration, by that of other reference quantities, having the same physical dimensions as the first, dimensionless numerical groupings are obtained. Thanks to dimensional analysis, it is possible to present experimental results no longer as a function of numerous parameters, but rather thanks to a few pure numbers with the further advantage of being able to extend validity to geometrically similar cases. Dimensional analysis therefore represents an excellent starting point for the study of a physical phenomenon and it is ideal preparation especially for its subsequent modelling; for these reasons it can be well used in debris flow investigations in order to get a dimensionless formulation of the equations that govern this phenomenon and to highlight the dimensionless parameters that influence their solutions. The aim is to succeed in explaining the functional dependence that exists between these groupings. The equations that govern the field of motion, which were obtained in Section 6.3 and following, can be posed in a dimensionless form, thus highlighting the parameters that influence their solutions. One of the techniques that can be used for getting a dimensionless formulation of a set of equations is that of analysing the differential equations that govern the problem and make them dimensionless by dividing the values of the quantities that appear there by those of suitable reference quantities having the same physical dimensions as the first [73]. The fundamental condition for the use of the method is that of considering geometrically similar physical situations, thus permitting the generalisation of the equations that represent the geometrical border, which is therefore identified by the shape of a family of surfaces and by a characteristic dimension [77]. The formulation of a dimensionless form of the motion equations fits therefore in the dimensional analysis of a certain physical phenomenon, as the aim is always to obtain formulae that explicitly contain dimensionless groupings with the reference quantities, thus consenting a wider use of the experimental results.

Therefore, by introducing U and L that represent a characteristic velocity and length of the phenomenon in question, respectively, the continuity eqn (6.2), becomes:

$$\frac{L}{U}\left(\frac{\partial u}{\partial x}+\frac{\partial v}{\partial z}\right)=0\cdot\frac{L}{U} \quad \Rightarrow \quad \left(\frac{\partial u'}{\partial x'}+\frac{\partial v'}{\partial z'}\right)=0\,, \tag{6.41}$$

where $u'=u/U$, $v'=v/U$, $x'=x/L$ and $z'=z/L$ are dimensionless quantities. In the case of a debris flow, for L, one could introduce, for example, the depth of the flow, which generally varies from 10 cm (in the case of laboratory experiments) to 1 m (in the case of medium-sized natural debris-flows), alternatively one could choose the width of the channel, which in turn could vary from 1 m (for laboratory experiments) and 10 m (in the case of medium-sized natural debris-flows; U could be estimated using $\sqrt{gL}$ [78]. In the same way, from the momentum equation, eqn (6.6), in the x direction, one obtains:

$$\frac{L}{U^2}\left(u\frac{\partial u}{\partial x}+v\frac{\partial u}{\partial z}\right)=\frac{L}{U^2}\cdot\frac{1}{\rho}\left[-\frac{\partial p}{\partial x}+\frac{L^2}{L^2}\left(\frac{\partial \tau_{zx}}{\partial z}+v\frac{\partial \tau_{xx}}{\partial x}\right)\right] \tag{6.42}$$

$$\Rightarrow$$

$$\left(u'\frac{\partial u'}{\partial x'}+v'\frac{\partial u'}{\partial z'}\right)=\left[-\frac{\partial p'}{\partial x'}+\frac{L}{U^2}\cdot\frac{1}{\rho}\frac{L^2}{L^2}\left(\frac{\partial\tau_{zx}}{\partial z}+v\frac{\partial\tau_{xx}}{\partial x}\right)\right],$$

in which one introduces $p'=p/(\rho U^2)$. At this point, we must point out the differences in studying the field of motion of a Newtonian fluid and of a non-Newtonian system, which is the case for a debris flow. Below, we will obtain two characteristic dimensionless groupings in the study of the motion field for Newtonian fluids and for non-Newtonian systems, respectively, evaluating as a function of them the stress exerted by the fluid itself on the reference surface.

6.6.1 Comparisons between Newtonian and non-Newtonian flows

In the case of a Newtonian fluid, the stresses acting have a viscous nature only and therefore by definition $\tau_{zx}=\mu\frac{\partial u}{\partial z}$ and $\tau_{xx}=\mu\frac{\partial u}{\partial x}$, which substituted into eqn (6.42) give:

$$\left(u'\frac{\partial u'}{\partial x'}+v'\frac{\partial u'}{\partial z'}\right)=\left[-\frac{\partial p'}{\partial x'}+\frac{L}{U^2}\cdot\frac{\mu}{\rho}\left(\frac{L^2}{L^2}\cdot\frac{\partial^2 u}{\partial z^2}+\frac{L^2}{L^2}\cdot\frac{\partial^2 u}{\partial x^2}\right)\right]$$

$$\Rightarrow \tag{6.43}$$

$$\left(u'\frac{\partial u'}{\partial x'}+v'\frac{\partial u'}{\partial z'}\right)=\left[-\frac{\partial p'}{\partial x'}+\frac{1}{N_{RE}}\left(\frac{\partial^2 u'}{\partial z'^2}+\frac{L^2}{L^2}\cdot\frac{\partial^2 u'}{\partial x'^2}\right)\right],$$

in which $N_{RE}=(\rho UL/\mu)$ is the Reynolds number. The physical significance of the Reynolds number is demonstrated by an examination of the momentum equation, eqn (6.6), by assuming the pressure gradient is null; in fact, the forces of inertia that represent the first member of eqn (6.6) have dimensions equal to U^2/L, in fact it is possible to make it dimensionless by multiplying and dividing by this quantity, whereas the viscous forces, which represent the second member, have dimensions equal to $\mu U/(\rho L^2)$. In short, their ratio represents the Reynolds number itself:

$$\frac{U^2/L}{\mu U/\rho L}=\frac{\rho UL}{\mu}=N_{RE}, \tag{6.44}$$

which allows us to state that for low number values (laminar regime) viscous forces are dominant, whereas for high values (turbulent regime) inertia forces prevail. This demonstrates the importance and the considerable potential that this dimensionless group offers in the physical description and use of experimental results of Newtonian fluids, but this is not all, it also allows us to claim that in a similar way to the Reynolds number for Netwonian fluids, there must be one or more dimensionless groupings characterising the hyperconcentrated flows, such as water–sediment mixtures [79]. As Newtonian fluids are characterised by viscous forces, flows with high granular material concentrations generate stresses due to the interaction between the grains thus causing non-Newtonian behaviour; dispersive collisional stresses are therefore central to these hyperconcentrated water–sediment mixtures, as highlighted in debris-flow phenomenology (Chapter 3). Therefore, by following this analogy, we recall the momentum balance equation, eqn (6.6), in the x direction for uniform, steady, and bidimensional flow of a water–sediment mixture, in which $\rho = \rho_s C + \rho_f(1-C)$ is the density of the mixture, C the solid concentration by volume, ρ_s and ρ_f the density of the granules and the interstitial fluid, respectively, it is also assumed that the stresses are exclusively caused by the collisional effect, thus neglecting initially the turbulence of the interstitial fluid. Many expressions have been proposed for the collisional stresses in solid–liquid mixtures (Bagnold, [7]; Savage et al., [53]; Hashimoto, [79]), however, generally speaking, almost all can be formulated using a simplified form, which is shown below for the bidimensional case:

$$\tau_{zx} = \alpha_{zx}\rho_s d^2 f(C)\left(\frac{\partial u}{\partial z}\right)^2$$
$$\tau_{xx} = \alpha_{xx}\rho_s d^2 f(C)\left(\frac{\partial u}{\partial z}\right)^2, \tag{6.45}$$

where τ_{zx} is the stress exerted in the x direction on a surface orthogonal to the z axis, τ_{xx} the stress exerted in the x direction on a surface orthogonal to the x axis, α_{zx} and α_{xx} are suitable numerical coefficients to be introduced, respectively, in the expressions of τ_{zx} and τ_{xx}, d is grain diameter and $f(C)$ a function that increases with the volumetric solid concentration; for the latter a suitable expression could be [79]:

$$f(C) = \frac{(C/C^*)^2}{1-C/C^*}, \tag{6.46}$$

in which C^* indicates the maximum solid concentration by volume possible for the mixture considered.

In the same way that the forces of inertia have dimensions equal to $\rho U^2/L$, as previously demonstrated, collisional forces must have dimensions equal to $\rho_s d^2 f(C) U^2/L^3$ by necessary dimensional homogeneity as we can see in eqn (6.45). Therefore the ratio between these terms gives:

$$\frac{\rho U^2/L}{\rho_s d^2 f(C) U^2/L^3} = \frac{\rho}{\rho_s f(C)}\left(\frac{L}{d}\right)^2. \tag{6.47}$$

This relationship illustrates the importance of the parameter L/d, which is the relative depth of the flow in the case in which one chooses the depth of the flow h as the characteristic quantity of length, and C, the solid concentration by volume. In short $(L/d)\sqrt{\rho/\rho_s f(C)}$ constitutes a characteristic dimensionless grouping for hyperconcentrated water–sediment flows: small values of this grouping, which are obtained for high C values and low L/d values, result in the prevalence of the term linked to the intergranular stresses in relation to the inertial term, whereas large values of this parameter, which are obtained with low C values and high L/d values, cause the prevalence of the inertial term on intergranular stresses, this could mean that turbulent stress, which has been neglected here, could become dominant. Certain experimental tests [79] suggest that for $(h/d)<20\approx30$ intergranular forces prevail, whereas for $(h/d)>100$ inertial forces prevail, thus providing an intermediate interval in which both effects are present simultaneously.

In the case of a Newtonian fluid the balance equation, eqn (6.7), of momentum in the z direction, on the basis of dimensional considerations of the terms that appear therein, let us to observe that the distribution of pressure can generally be traced back to the hydrostatic distribution $\frac{\partial p}{\partial z} = -\rho g$ and therefore the pressure gradient is as a datum of the problem linked to the geometric configuration and to the velocity distribution of the developed motion [73]. Bearing this in mind and considering eqns (6.41) and (6.43), it can be concluded that for a given geometry, the dimensionless form of the motion field and the corresponding friction coefficient C_f in the case of a Newtonian fluid depend exclusively on the Reynolds number, in other words $C_f=C_f(N_{RE})$, and the shear stress exerted on the wall τ_s by the fluid above can be written as:

$$\tau_s = (\tau_{zx})_{z=0} = \mu \left.\frac{\partial u}{\partial z}\right|_{z=0} = C_f \rho \frac{U^2}{2}. \tag{6.48}$$

Therefore knowing the friction coefficient, which can be calculated from the correlations with the Reynolds number available in the literature, permits the direct calculation of shear stress at the wall. If we then wish to estimate the shear stress at the channel bed, indicated by τ_0, for a uniform, steady and bidimensional flow of a water–sediment mixture, in eqn (6.47) we must introduce the height z measured from the bed as the characteristic length L; by assuming the concentration profile to be uniform for simplicity, the stress distribution can be expressed by $\tau_0 = \left(1 - \frac{z}{h}\right)$. At the channel bed it is possible to neglect z/h in relation to 1 and therefore pose $\tau \approx \tau_0$, it can also be assumed that in this region, known as the *granular sublayer* [79], the inertial effect that can be associated with turbulence is negligible in relation to the intergranular collisions; this leads to:

$$\tau_0 = \alpha_{zx} \rho_s d^2 f(C) \left(\frac{du}{dz} \right)^2 . \tag{6.49}$$

By introducing a characteristic velocity representative of the stress τ_0 on the bed, known as shear velocity and indicated with u^*, by dimensional homogeneity, it must follow that $\tau_0 = \rho u^{*2}$. If we then integrate eqn (6.49) with the no-slip hypothesis at the bed ($u_{(z=0)}=0$), we can obtain, as an initial approximation, a linear velocity profile:

$$\frac{u}{u^*} = \frac{\xi}{\sqrt{\alpha_{zx}}} \cdot \sqrt{\frac{\rho}{\rho_s f(C)}} \left(\frac{z}{d} \right), \tag{6.50}$$

where a corrective parameter ξ has been introduced having hypothesised a uniform distribution of concentration. From the previous equation, it is obvious that the dimensionless velocity profile is a function of $(L/d)\sqrt{\rho/\rho_s f(C)}$, the dimensionless grouping previously introduced, which represents the ratio between the inertial and collisional effects, as we can also write:

$$\frac{u}{u^*} = f\left(\frac{z}{d} \cdot \sqrt{\frac{\rho}{\rho_s f(C)}} \right). \tag{6.51}$$

This relationship is somewhat similar to the “wall law” valid for Newtonian fluids. It should be pointed out that relationship (6.51) is only a partial result, as it is subject to considerable simplifications, including having considered the effect of collisional stress only, when debris flows include a wide range of types of stress (static, cohesive, turbulent, and viscous are the main ones) and therefore at the current state it is difficult to obtain a formulation of the motion field, using an explicit dependency on the dimensionless groupings (of which there are also many) characteristic of the phenomenon, which reflects that eqn (6.43) is valid for Newtonian fluids. Despite the limits in the dimensionless formulation of motion equations, by permitting the definition of dimensionless parameters through the analysis of the variables that condition the system, dimensional analysis represents a valid instrument for improving the understanding of these very complex natural phenomena and in particular, for identifying clearly which are the processes responsible for momentum transport and the mixture’s stress state.

6.7 A debris-flow classification based on dimensional analysis

A ternary phase diagram has been proposed for the systematic classification of the various types of debris flows that can occur in nature [41]. This classification is based on the Bagnold number and Reynolds number and on the relative depth of the flow *h*/*d*. Bagnold's number represents the ratio between the collisional effect and the viscous effect, and Reynolds number the ratio between the inertial effect and the viscous effect, while the relative depth is the ratio between the depth of the flow and granule diameter.

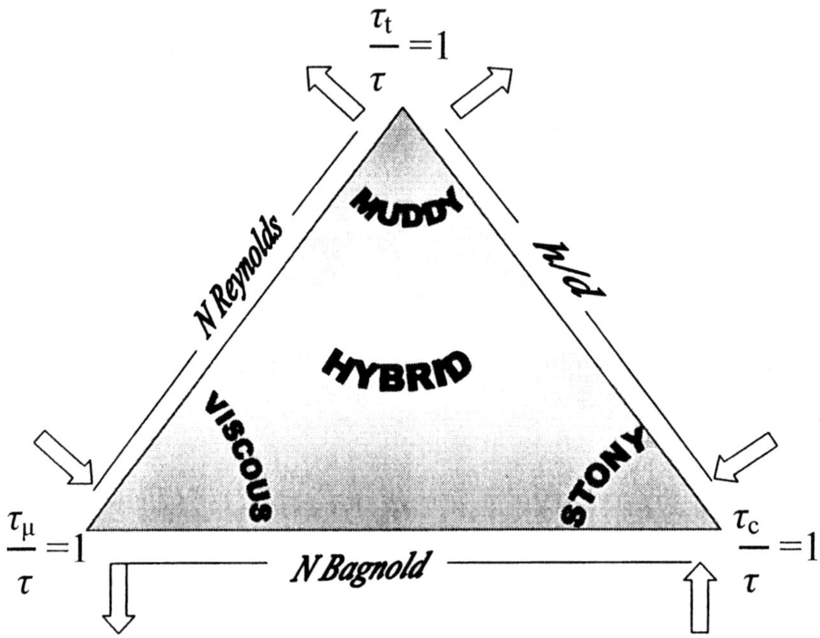

Figure 6.10: Ternary phase diagram proposed by Takahashi [41] for the classification of various types of debris flows (with $0.3<C<0.5$), using dimensionless grouping. In the diagram τ is overall shear stress, whereas τ_c, τ_μ, τ_t are the collisional, viscous and turbulent components, respectively.

The three fundamental parameters of this classification method quantify, in order, the effect of the continuous and intense contacts between particles, the mixing due to turbulence and to characteristic fluctuation motion of the particles and the effect of the viscoplasticity. In this classification values of the solid concentration between 0.3 and 0.5 are considered [41]. In fact, if the concentration exceeds 0.5, the prolonged contacts between the particles cause a quasistatic motion condition in which the dominance of Mohr–Coulomb's static stresses can be observed, and, if this value drops below 0.3, one has a particularly dilute flow, which Takahashi [41] terms *immature debris-flow*. In this case, the dispersion of the larger particles does not cover the entire depth of the flow but rather stratification occurs, for which on the bed, a highly collisional mixture, involving the coarser particles, concentrates above which a turbulent flow of water and fine particles flows.

The tips of the diagram bear the types of debris flows with clearly defined characteristics, whereas the somewhat vast central area corresponds to a set of numerous types of behaviour grouped together as *hybrid debris-flows*.

- *Stony debris flows*: these are relatively shallow flows (relative depth less than 20) with very high Bagnold's numbers, of which the particle-dispersion mechanism throughout the depth of the debris-flow is dominated by collisional stresses. These flows are characterised by high solid particle concentrations and dimensions, whereas the fine content in the interstitial fluid is rather low. During descent, the movement and consequential accumulation of large boulders towards the head of the wave take place (see Section 3.4.4), a phenomenon that is negligible in the other two types of debris-flow.
- *Muddy debris flows*: relative depth and Reynolds number correspondent are both consistent, from this we can deduce that the momentum-transport mechanism is dominated by turbulent mixing and therefore, when the flow encounters an obstacle, it violently covers it with large quantities of mud. The particles dispersed in the fluid are small, but cohesion between them is not yet significant enough to form an overall viscous mixture.
- *Viscous debris flows*: when the Bagnold number and the corresponding Reynolds number both become negligible, the viscosity of the interstitial fluid prevails within the flow, only the head still has considerable turbulence, but the rest of the flow moves in a laminar way.
- *Inertial debris flow*: the class of hybrid debris-flows, which range from muddy to stony type along the h/d axis, are termed by Takahashi inertial debris-flow. They consist in a lower layer, dominated by the brief inelastic collisions between the larger grains and by the contact friction between them, and an upper one constituted by a turbulent suspension in which the cohesive particle content in the interstitial fluid is minimal; the relative extension of the two layers depend naturally on the concentration and particle diameter. The fundamental difference between an "immature" and an inertial or hybrid debris-flow is that in the former characteristic particle suspension, which is, on the contrary, a peculiar characteristic of the others, is not observed in the upper layer, as it is constituted by a simple flow of water.

Chapter 7

Debris-flow disaster mitigation

7.1 Introduction

The terms resource and natural disaster are a perfect reflection of mankind's perception of natural systems. When man recognises a natural system as useful, it becomes a resource, whereas if it is potentially harmful, it is seen as a danger. In the absence of human settlement, however, there are neither resources nor dangers, but merely natural systems [27]. Debris flows are mainly the result of the natural processes of erosion and sediment transport, which cause disasters when they interact with human settlements. The mitigation of such phenomena could therefore be obtained by reducing this interaction, by altering the natural system, the human system, or both considering social impact on the one hand and environmental impact on the other. (Figure 7.1).

The engineering approach attempts to tackle the natural system as the cause of the problem, trying to reduce potential interaction by means of structural measures. However, engineering works aimed at debris-flow mitigation (dykes, dams and deposition basins) are not currently entirely reliable, in fact, although the planning of collection basins may appear simple, it requires a complete knowledge of the phenomenon, such as the exact forecast of flow volumes. Given the complexity of natural systems, direct intervention on debris-flows using structural measures requires parallel intervention on the human system, which could involve the transfer of people and resources from hazard areas, either temporarily, using warning and evacuation systems, or permanently with the regulations of hazard areas. Two important aspects make measures for debris-flow hazard mitigation economically and technically sustainable: prevention and protection [1]. There are two approaches to the protection problem, involving active and passive intervention measures respectively [80]. Protection systems based on a structural approach constitute active measures as they involve altering

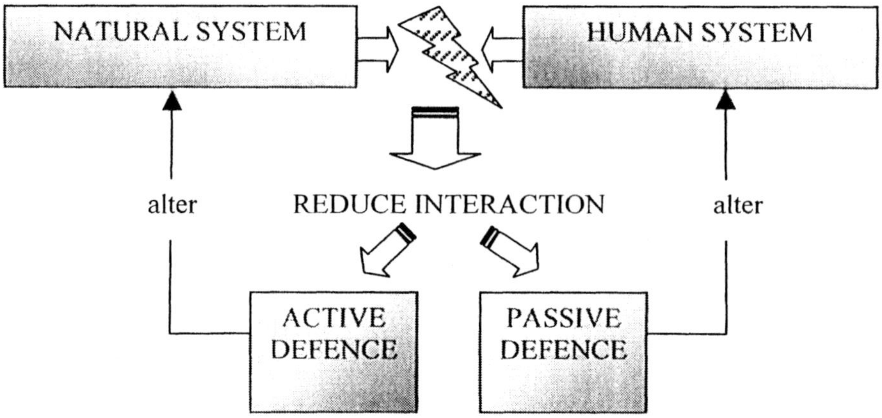

Figure 7.1: The diagram shows how in order to reduce the interaction between the natural system and human systems, alterations must be made to either the natural system, the human system or both, considering social impact on the one hand and environmental impact on the other.

the natural area that is potentially at risk from debris-flows by constructing suitable structures aimed at minimising the damage they cause; non structural mitigation measures, are passive in that they focus on regulations of the area and installing alarm systems. In short, a complex system of countermeasures for tackling the debris-flow problem using prevention and protection must be characterised by a series of basic measures that formulate the *hardware*, and by a number of functioning programmes that can be considered its *software* [81].

7.2 Nonstructural measures

Passive measures involve all those countermeasures aimed at avoiding or preventing the use of areas subject to debris-flow transit and/or deposition, either permanently or in immediate hazard conditions. They do not alter the natural system, as they do not change the characteristics of the event, but merely reduce interaction with human activity [27]. Passive measures are sometimes the only ones applicable due to the limited intervention times of sudden events, or the excessive cost of structural work. These measures do not always prove reliable, especially in the case of automatic alerting devices where false alarms can occur, thus increasing the system's management costs and creating situations of collective confusion. The mapping of debris-flow hazard areas is also essential. Debris-flow consequences derive from the direct impact with the flow or with the succession of the phenomenon's stabilisation waves that can continue for as much as several days; the impact is most violent during the initial stages of the process, which is characterised by a high kinetic energy content, whereas lesser consequences occur in the fan areas, where deposition takes place. It is therefore necessary to define the corresponding trigger and deposition areas for a given topography and a range of different restrictive measures must be adopted [1]. In evaluating risk, which is above all fundamental for areas where the impact on human settlements would be considerable, we can use the statistical method, by which risk is expressed by means of a cumulative function linked to various events, or the probabilistic method that introduces the frequency of event occurrence. One equation to evaluate the specific risk, calculated using the former method, could be $SR = E \times H \times V$, in which E represents the economic value of the threatened element, H is the probability of event occurrence calculated for a given timeframe and V is vulnerability, which represents the fraction of the element that is in potential danger and that could be involved in a specific event; for example, for an urban area E=100 billion \$, H=0.001 for a period of ten years and V=0.01, then SR=1 million \$.

Territorial mapping can also be complemented by the use of alarm signals triggered by the event and the subsequent evacuation of the area. Given that evacuation in emergency conditions requires hours rather than minutes, the alarm signals triggered by the event do not provide sufficient time to complete evacuation; in such conditions, it is essential to have accurate information on the exactly available warning time and on the time required for evacuation, in order to understand whether an alarm system triggered by the event, combined with the evacuation of the area provides the most suitable mitigation strategy. If not, the only alternative is to resort to monitoring the potential debris-flow basin and weather forecasts, which has the advantage of longer warning timeframes and the disadvantage that it is not a particularly accurate system and can lead to undesirable false alarms [27]. It is within this context that communication and information media play a fundamental role in supporting the credibility of alarm systems, by persuading public opinion of the difficulties in forecasting debris-flows and keeping sensitivity to general perception of natural risk high.

The main passive intervention forms can be summarised as follows [82]:

- systems based on monitoring, morphological changes to the territory and on the study of past phenomena;
- advance-warning systems, based on monitoring environmental conditions, which, using controlled feedback measurement chains, foresee the possible occurrence of an event if the values of the quantities measured exceed the corresponding threshold values;
- alarm systems designed to activate as soon as a new event takes place, by simultaneously signalling and monitoring the phenomenon in real time;
- systems signalling the end of the event, useful in reducing further damage that could occur from the use of structures that have become unsafe.

7.3 Structural measures

Active countermeasures aim to reduce the supply of debris to the collection basin and to the debris-flow in progress, using structural works that reinforce the stability of the hillside. Measures that act directly on the phenomenon itself are based on drainage systems that aid the removal of the interstitial fluid from the debris-flow deposits; reforestation and slope consolidation; indirect measures, on the other hand, include dykes, dams and deposition basins that control and can modify the flow [2]. Such structures aim to reduce the debris-flow velocity, bed-erosion capacity and discharge and deviate flow direction. Planning criteria for such works often depend on work conditions and experience, using the hypothesis of a locally uniform flow and kinematic flow parameters typical of the phenomenon: velocity, maximum discharge or hydrogram and volume [2]. The definition of structural interventions should also take into consideration the rheological behaviour of the flow; as debris-flows can involve the movement of considerably sized material at high velocities along particularly steep slopes and being characterised by a particularly destructive head, mitigation measures must attempt to halt the coarser fraction of the debris-flow and the dampening of the forces with destructive impact by draining the interstitial fluid.

Debris-flow defence works can therefore be summarised as follows [81]:

1. *Works aimed at limiting the frequency of phenomenon occurrence*: these measures prevent the movement of debris along slopes and consist mainly in stabilisation and consolidation structures, dykes and naturalistic engineering work.
2. *Debris-interception devices*: these prevent the movement of debris along the channel bed; they capture part of it in order to gradually reduce the solid discharge of the flow and hinder the large boulders present in the head; they consist mainly of dams for stopping the head wave and withholding the larger clasts in the upper part of the basin.
3. *Flow-direction-deviation devices*: these devices direct the flow towards areas where human impact is lower by means of a branching system of artificial training dykes spaced out by dams and deposition basins in order to drain the turbulent fluid. In such conditions, the flow loses a substantial part of its mobility. Dykes should preferably be straight and should avoid sudden variations of the longitudinal profile, given that debris flows generally proceed in a straight line.
4. *Lateral flow-containing banks*: these structures control the direction of the flow and prevent overtopping by means of sufficiently high walls.
5. *Reforested areas for dispersing debris*: when used in combination with the above-mentioned measure, reforested areas contribute to controlling the quantity of deposits in the alluvial areas, thus acting as buffers between the descending flow and the target to be protected.

6. *Accumulation basins for solid material transported:* these measures encourage the stoppage or slowdown of the flow through a gradual dissipation of energy.

The strategy adopted is that of installing combinations of the abovementioned structures along the potential debris-flow path, considering the topographical conditions, structural flexibility and costs (Figure 7.2).

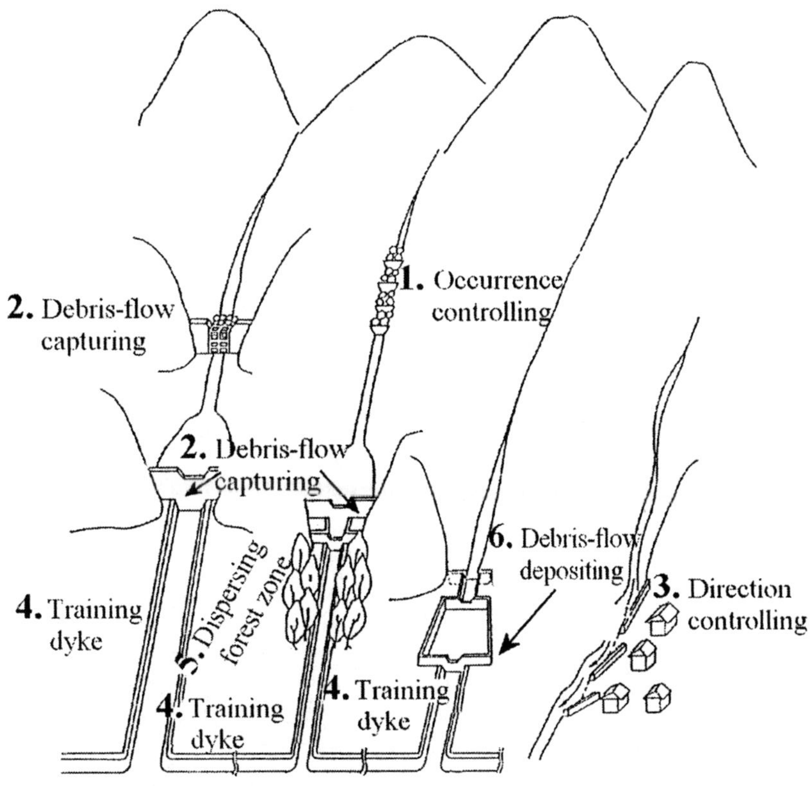

Figure 7.2: The diagram shows a typical example of intervention by means of the combination of various types of structural measures used coherently with the particular characteristics of the occurrence, propagation, and deposition areas of the debris flow [81].

Debris-flow phenomenology demonstrates that flows develop mainly along three areas: the debris-collection basin, the transport area and the deposition area. As mentioned previously with regards to territorial mapping, correct identification of these areas for a given topography allows the use and installation of customised countermeasures, increasing the efficiency of the mitigation system as a whole. Preventing the movement of deposits on the channel bed is of primary importance in attempts to control debris-flow triggering in its main development area. An area could be a potential debris-flow source if the slopes are steep or if it has superficial ridges similar to scars, which point to the occurrence of past events. In this case, attempts are made to consolidate the slope and make it safe using civil engineering work, by increasing the stability of the slope with retainer foundations built in cement-reinforced stone as well as installing superficial or buried drainage systems [1]. In flow-transit areas that directly or indirectly interfere with human settlements, bridges or tunnels, deviation devices and stoppage walls are employed, aimed at dampening the impact of the flow and its velocity. Lastly, artificial accumulation basins, which will be emptied once the event has terminated are built in deposit areas [80].

Although structural events do not directly alter the human system, they cannot be considered as efficient long-term protection systems, and they are not capable of countering the destructive impact of debris flows, in fact, the reliability of these structures is conditioned by full understanding of the phenomenology connected to debris-flows that have taken place in the past, as well as possible future events.

7.4 Development prospects

Immediate debris-flow hazard-mitigation measures must include a meticulous empirical description of the geomorphology of threatened areas and their rational planning, integrated when necessary with suitable alarm–evacuation systems that permit continued habitation of these areas and therefore reduce the inevitable social and economic impact that any measure aimed at reducing interaction between the natural and human systems causes when it intervenes directly on the latter. However, despite the fact that structural measures aimed at directly altering debris-flow behaviour do not interfere with human settlements, they do not eliminate the risk of serious consequences due to the destructive impact of the phenomenon nor do they represent a feasible long-term solution. Generally speaking, the behaviour of a natural system is nonlinear as the system includes many mutually interacting components and it is also probable that each component is in itself nonlinear and the behaviour of the entire system is chaotic. This makes a natural system unpredictable, at least in the immediate future [27]. Debris flows pertain to this type of system, because of the many components that interact with one another: nonlinear rheological behaviour that varies in space, nonlinear variation in interstitial pressure and interaction between the solid and liquid phases. Despite the fact that the behaviour of a natural system is intrinsically unpredictable, it is not, however, completely random; at a certain scale, it is possible to observe certain behavioural aspects that dominate the individual local processes and that can be described empirically and independently. It is hoped that a vast, international-scale programme can be developed for the systematic collection of data on the occurrence and characteristics of debris-flows in a variety of geological and climatic situations, aimed at standardising the acquisition of data and the reporting system [27]. This database could provide tangible back-up in selecting the most suitable strategies for debris-flow risk mitigation and help the development of realistic physical forecasting models. With regards to this prospect in particular, debris-flow rheology requires continuous research activity and, as the effectiveness of active and passive defence mechanisms depends to a large extent on the knowledge and understanding of the natural system, we have preferred to put a complete analysis of the phenomenology and dynamic behaviour of debris-flows before the question of intervention techniques, which have been only summarily treated in this chapter and for which we refer the reader to specific literature (see [5, 27, 80–82]).

Chapter 8

Concluding remarks

8.1 Summary

The recognised need to improve knowledge on the mechanics of solid–liquid flows, highlighted by a critical analysis of the current international state of the art, prompted the study carried out in this book on debris-flows process mechanics, starting from a phenomenological examination.

With regards to this approach, this work aims to provide an applicable methodological insight for investigations into the mechanics of sediment-transport processes in debris flows, through a methodology able to give an accurate representation of the phenomenology. The book tackled the following subjects (Figure 8.1): from phenomenological investigations, which highlighted the main characteristics of debris-flow mechanics (dilatancy, internal friction and cohesion, fluidification, segregation, sediment-transport mechanisms) which should therefore be suitably incorporated into any attempt at modelling, and from rheological studies, which, by shedding light on the properties of the mixture, aim to give a macroscopic description using experimental investigations, it came to a critical analysis of the various modelling approaches available in the literature.

The main results obtained concern the understanding of certain key aspects of the mechanics of solid–liquid flows and the consequential application of mathematical modelling.

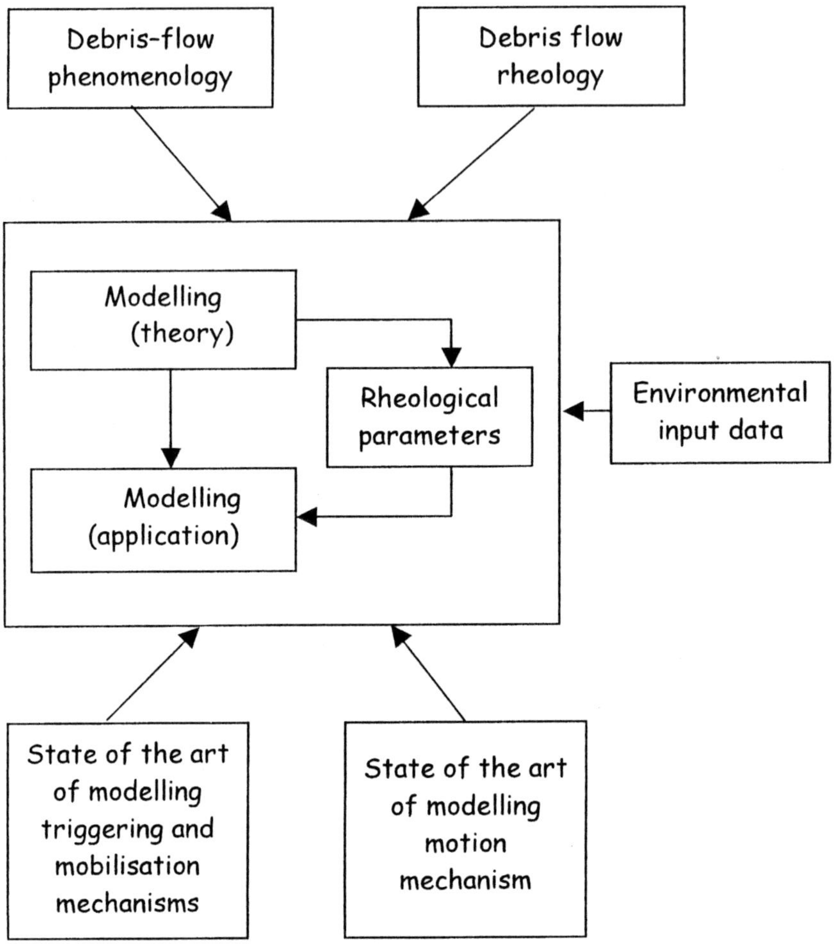

Figure 8.1: The chart, shows the individual phases of the debris-flow study approach adopted in this book.

8.2 Considerations on the mechanics of solid–liquid flows

The main difficulties that hinder the study of debris-flow mechanics are connected to the multidisciplinary approach that a full comprehension of these phenomena require. In fact, certain key concepts of debris-flows study are borrowed from different disciplines (geomorphology, geology, hydrology, fluid mechanics, rock mechanics, and soil mechanics); this transposition of contents (see the saturation question raised in Section 3.7) must, however, be carefully checked in order to describe the mechanics of debris-flows in a satisfactory way that also takes into due account the peculiarities that characterise the process. In particular, the recognition that the complex interactions between the solid and liquid phase univocally characterise a debris-flow, thus contributing to distinguishing it from other similar transport processes, leads to the conclusion that a debris-flow effectively constitutes a "*sui generis phenomenon*" from a mechanical point of view, due to both the particular conditions in which it occurs and the particular constitutive characteristics of this type of two-phase system. The study of the various types of transport mechanics highlighted how within a hyperconcentrated flow, composed of solid material and water, several types of transport can occur, depending on the variation in sediment concentration and dimensions. Subsequent research could therefore focus on establishing a connection between various concentration intervals within the flow and the different types of stresses predominant in each thus identify, once and for all, an indicator for flow type, which would therefore allow the distinction between the various types of sediment transport. However, at the current state, the limit values of the solid concentration proposed for these intervals, if we can indeed speak of limit value, not only are unclear, but also appear somewhat arbitrary. Despite the fact that in nature a wide range of debris-flows can be observed due to the dominance of different sets of stresses acting within the mixture rather than others, which could suggest the use of different models according to the case, on the other hand, irrespective of its possible forms, the debris-flow phenomenon is characterised by the simultaneous presence of four aspects that characterise momentum transport in the mixture, in other words the viscous, collisional, turbulent and yield stress effects and it could therefore be said that, as it considers for definition the simultaneous presence of all four contributions, the visco plastic-collisional regime is the one that best represents debris-flow mechanics.

8.3 Hypotheses for future progress

The critical analysis contained in the previous chapters on the current state of the art of debris-flow modelling related to trigger mechanisms and the dynamics of motion made it possible to highlight the vast research areas that require further progress. Although the physical process would seem to be understood, certain aspects of the phenomenon's mechanics must be better investigated, and in particular the problem of the development of suitable constitutive equations and the study of the phenomenon dynamics, topics that have been dealt with in this book. The development of constitutive equations is an aspect of debris-flows that has received particular attention; most models are based on a rheological approach, which has the drawback of providing equations that require a large number of parameters, whose number and type depends on the complexity of the relationship, and whose definition is often problematic, if one considers the vast range of flow regimes that can occur. One alternative is that of developing and using simple models (see hydraulic approach in Section 5.4.8) to focus attention on one salient question of debris-flow modelling, the dynamic aspect. From this point of view the constitutive equations must be compatibly incorporated into the conservation equations in order to obtain a realistic representation of the phenomenon. This problem immediately leads to experimental studies on debris-flows: there is, as yet, relatively scarcity of experimental data, the only ones that consent effective verifications of the constitutive models proposed, in different flow situations, and the estimation of the rheological parameters they contain. It is also important to focus laboratory experiments on investigations that can shed light on the effective relationship between sediment concentration and flow parameters. The experimental point of view in debris-flow research, however, encounters considerable problems that are yet to be fully overcome, connected largely to the accuracy of measuring techniques and flow simulation in experimental tests. Lastly, field studies are probably the most difficult and costly study approach of debris-flows; the difficulties encountered are connected to their considerable complexity and the difficulty of direct observation. The exceptional and infrequent conditions in which debris-flows occur do not generally permit a sufficient number of observations for the same type of territorial reality to deduce the specific behavioural laws for that area. Reference to different territorial situations also highlights another problem: that of the homogeneity of data, given the substantial territorial peculiarity in which the phenomena occur. Besides, field data is essential in determining the quality of any mathematical model, as it is especially important for estimating velocity, discharge, concentration, yield stress, viscosity, and grain-size. This need requires the use of laboratory experimentation when the previous problems cannot be overcome and in certain cases, it is the only possible path to follow. However, the achievement of good agreement between theoretical and experimental, results does not justify indiscriminate extrapolations for the various territorial situations, which have very different boundary conditions from standard laboratory conditions. Assuming that the scientific research path cannot exclude an accurate observation and description of the phenomenon in question, without which the analysis of physical processes, that generate it, would become extremely artificial and uncertain, it is hoped that any attempt at improving the interpretations of the phenomenon involves critical comparison between the theoretical, experimental, and field approaches, as well as extensive osmosis process between the same approaches.

In concluding this work with a glance to the direction to be followed in future debris-flow in-depth studies, we will close with the point of view expressed by one of the subject's most illustrious international researchers, Prof. Pierre Y. Julien of the Colorado University, Civil Engineering Department:

"If in practice, insufficient information is available for constructing any type of model, this must not discourage us from undertaking studies on such complex questions as debris flows; in fact, it is sufficient to think of the devastating results caused by debris-flow events throughout the world, to

highlight how many steps research must still make towards full comprehension of the phenomenon. At the current time, what it is worth doing is attempting to develop an engineering technology that can help resolve real problems and develop structural solutions that can help save human lives.

Only by making progress in this direction can we claim to be making progress in debris flow studies."

References

[1] Mainali, A. & Rajaratnam, N., *Hydraulics of Debris Flows: a Review*, Tech. Rep. (WRE 91–2), Dept. Civ. Eng.., Univ. of Alberta, Edmonton, Alberta, Canada, 1991.

[2] Armanini, A., Control measures: Introduction, *Recent Developments on Debris Flows*, Aronne Armanini & Masanori Michiue (eds), Springer, pp. 163–165, 1997.

[3] Iverson, R. M., The physics of debris flows, *Reviews of Geophysics*, **35(3)**, pp. 245–296, 1997.

[4] Wan, Zhaohui, *Hyperconcentrated Flow*, Monograph Series of IAHR, Rotterdam, 290 pp., 1994.

[5] Armanini, A. & Gregoretti, C., Triggering of debris-flow by overland flow: a comparison between theoretical and experimental results, *Debris flow Hazards Mitigation: Mechanics, Prediction and Assessment*, Wieczorek & Naeser (eds), Balkema, Rotterdam, pp. 117–123, 2000.

[6] Citrini, D. & Noseda, G., *Idraulica*, Casa Editrice Ambrosiana, Milano, 1987.

[7] Bagnold, R. A., Experiments on a gravity–free dispersion of a large solid sphere in a Newtonian fluid under shear, *Proc. R. Soc. London*, Ser. A, **225**, pp. 49–63, 1954.

[8] Ghetti A., *Idraulica*, Edizioni Libreria Cortina, Padova, 1980.

[9] Johnson, A. M., *Physical Processes in Geology*, Freeman, Cooper & Company, pp. 431–571, 1970.

[10] Pierson, T. C. & Costa, J. E., A rheological classification of subaerial sediment–water flows, *Review in Engineering Geology*, American Geological Society, **VII**, pp. 1–12, 1987.

[11] Sharpe, C. F. S., *Landslides and Related Phenomenon*, Cooper Square Publications Inc., New York, 1938.

[12] Varnes, D. J., *Landslide Type and Process, in Landslide & Engineering Practice*, ed. E.B. Eckel, Highway Research Board, special Report 21, 1958.

[13] Chen, L. C., Generalized viscoplastic modelling of debris flow, *J. Hydraulic Eng.*, **114**, pp. 237–258, 1988.
[14] Takahashi, T., Debris flow, *Ann. Rev. Fluid Mechanics*, **13**, pp.57–77, 1981.
[15] O' Brien, J. S. & Julien, P. Y., Physical properties and mechanics of hyper-concentrated sediment flows, *Proc. Specialty Conference on Delineation of Landslides, Flash Flood & and Debris Flow Hazards in Utah*, pp. 260–279, 1985.
[16] Coussot, P., *Rhéologie des Laves Torrentielles*, Collection Etude, serie Montagne, n°5, 415 pp., 1992.
[17] Savage, S. B., The mechanics of rapid granular flows, *Advances in Applied Mechanics*, **24**, pp. 289–366, 1984.
[18] Iverson, R. M., A constitutive equation for mass-movement behaviour, *J. of Geology*, **93**, pp. 143–160, 1985.
[19] Beverage, J. P. & Culbertson, J. K., Hyperconcentrations of suspended sediment, *J. Hydraulics Division*, ASCE, **90**, n° HY6, pp. 117–126, 1964.
[20] Chen, L. C., Comprehensive review of debris flow modelling concepts in Japan, *Review in Engineering Geology*, American Geological Society, **VII**, pp. 13–29, 1987.
[21] Takahashi, T., *Debris Flow*, Monograph Series of IAHR: Rotterdam, 1991.
[22] Fei, X., Grain composition and flow properties of heavily concentred suspensions, *Proceedings, Second International Symposium on River Sedimentation*, pp. 307–308, 1983.
[23] Coussot, P., Laigle, D., Arattano, M., Deganutti, A. & Marchi, L., Direct determination of rheological characteristics of debris flow, *J. Hydraulic Eng.*, **124(8)**, pp. 865–868, 1998.
[24] Franzi, L., On the effect of clays on collisional stresses in debris flow, *Proceedings of Second IAHR Symposium on River, Coastal and Estuarine Morphodynamics*, Obihiro, Japan, 10–14 Sept., 2000.
[25] Bin, Y. & Huilin, Z., Direct determination of rheological characteristics of debris flow, *J. Hydraulic Engineering*, **126**, pp. 158–159, 2000.
[26] Deganutti, A. M. & Scotton, P., Yield stress of granular material, *Proc. First International Conference Debris-flow Hazards Mitigation: Mechanics, Prediction and Assessment*, New York, ASCE, pp. 270–277, 1997.
[27] Davies, T.R., Using hydro science and hydro technical engineering to reduce debris-flow hazards, *Proc. First International Conference Debris-flow Hazards Mitigation: Mechanics, Prediction and Assessment*, New York, ASCE, pp. 787–809, 1997.
[28] O' Brien, J. S. & Julien, P. Y., Laboratory analysis of mudflow properties, *J. Hydraulic Engineering*, **114**, pp. 877–887, 1988.
[29] Locat, J., Normalized rheological behaviour of fine muds and their flow properties in a pseudo plastic regime, *Proc. First International Conference*

Debris–flow Hazards Mitigation: Mechanics, Prediction and Assessment, New York, ASCE, pp. 260–269, 1997.

[30] Costa, J. E., Physical geomorphology of debris flows, *Developments and Applications of Geomorphology*, eds J. E. Costa & P. J. Fleisher, pp. 268–317, 1984.

[31] Znamensky, D. & Gramani, M. F., Debris-flow grain-size analysis, *Debris-flow Hazards Mitigation: Mechanics, Prediction and Assessment*, Wieczorek & Naeser (eds), Balkema, Rotterdam, pp. 537–546, 2000.

[32] Julien, P. Y. & O'Brien, J. S., Selected notes on debris flow dynamics, *Recent Developments on Debris-Flows*, Aronne Armanini & Masanori Michiue (eds), Springer, pp. 144–162, 1997.

[33] Takahashi, T., Dynamics of the inertial and viscous debris flows, *Recent developments on Debris-Flows*, Aronne Armanini & Masanori Michiue (eds), Springer, pp. 117–143, 1997.

[34] Pierson, T. C., Dominant particle support mechanism in debris flow at Mt. Thomas, New Zealand, and implications for flow mobility, *Sedimentology*, **28**, pp. 49–60, 1981.

[35] Campbell, C. S., Rapid granular flows, *Annual Review of Fluid Mechanics*, **22**, pp. 57–92, 1990.

[36] Vanoni, V. A., *Sedimentation Engineering*, Task Committee for the Preparation of the Manual on Sedimentation, ASCE, 1975.

[37] Einstein, A., Eine neue bestimmung der molekuldimensionen, *Annalen der Physik*, Leipzig **(4) 19**, 1906.

[38] Bozhinskiy, A. N. & Nazarov, A. N., Two-phase model of debris-flow, *Debris-flow Hazards Mitigation: Mechanics, Prediction and Assessment*, Wieczorek & Naeser (eds), Balkema, Rotterdam, pp. 263–268, 2000.

[39] Simoni, A., *Innesco e mobilizzazione di debris-flow: il bacino sperimentale di Acquabona (BL)*, Dottorato di Ricerca in Scienze della Terra XI ciclo, (per gentile concessione del Prof. Elmi, Facoltà di Geologia, Univ. Bologna, Italia), 1998.

[40] Hutter, K., Svendsen, B. & Rickenmann, D., Debris flow modelling: a review, *Continuum Mechanics and Thermodynamics*, **8**, pp. 1–35, 1996.

[41] Takahashi, T., Initiation and flow of various types of debris-flow, *Debris-flow Hazards Mitigation: Mechanics, Prediction and Assessment*, Wieczorek & Naeser (eds), Balkema, Rotterdam, pp. 15–25, 2000.

[42] Davies, T.R.H., Debris flow surges-a laboratory investigations, *Mitteilungen der Versuchsanstalt fur Wasserbau*, Hydrologie und Glaziologie, Zurich, 1988.

[43] Ogawa, S., Multitemperature theory of granular material, *Proc. U.S. Japan Seminar on the Continuum Mechanical and Statistical Approaches in the Mechanics of Granular Material*, S. C. Cowin and M. Satake eds., Gakujutsu Bunken Fukyukai, Tokyo, Japan, pp. 208–217, 1978.

[44] Simoni, A. et al., Debris flows in Dolomites: experimental data from a monitoring system, *Debris-flow Hazards Mitigation: Mechanics,*

Prediction and Assessment, Wieczorek & Naeser (eds), 2000 Balkema, Rotterdam, pp. 283–293, 2000.

[45] Cavazza, L., *Fisica del Terreno Agrario*, UTET, 1981.

[46] Shen, H. H. & Ackermann, N. L., Constitutive relations for fluid–solid mixtures, *J. Eng. Mech. Div.*, ASCE, **108**, pp. 748–763, 1982.

[47] Terzaghi, K., *Theoretical Soil Mechanics*, John Wiley, 1943.

[48] Shields, A., Anwendung der Ahnlichkeits-Mechanik und der Turublenzforschung auf die Geschiebebewegung, *Preussische Versuchsantalt fur wasserbau und schiffbau*, **26**, pp. 5–24, 1936.

[49] Jan, C-D. & Shen, H. W., Review Dynamic Modelling of Debris Flows, *Recent Developments on Debris-Flows*, Aronne Armanini & Masanori Michiue (eds), Springer, pp. 93–116, 1997.

[50] Pasquarell, G. C. et al., Collisional stress in granular flows: Bagnold revisited, *J. Engineering Mechanics*, **114**, pp. 49–64, 1988.

[51] Goodman, M. A. & Cowin, S. C., A continuum theory for granular materials, *Archive for Rational Mechanics and Analysis*, 44, pp. 249–266, 1972.

[52] McTigue, D. F., A non-linear constitutive model for granular materials: application to gravity flow, *J. Applied Mechanics*, **49**, pp. 291–296, 1982.

[53] Savage, S. B. & Sayed, M., Stresses developed by dry cohesionless granular materials sheared in an annular shear cell, *J. Fluid Mechanics*, **142**, pp. 391–430, 1984.

[54] Hungr, O. & Morgenstern, N., Experiments on the flow behaviour of granular materials at high velocity in an open channel, *Geotechnique*, **34(3)**, pp. 405–413, 1984.

[55] Savage, S. B., Gravity flow of cohesionless granular materials in chutes and channels, *J. Fluid Mech.*, **92**, pp. 53–96, 1979.

[56] Johnson, P. C., Nott, P. & Jackson, R., Frictional–Collisional equations of motion for particulate flows and their application to chutes, *J. Fluid Mechanics*, **210**, pp. 510–535, 1990.

[57] Haff, P. K., Grain flow as a fluid–mechanical phenomenon, *J. Fluid Mechanics*, **134**, pp. 401–430, 1983.

[58] Savage, S. B. & McKeown, S., Shear stresses developed during rapid shear of concentrated suspensions of large spherical particles between concentric cylinders, *J. Fluid Mechanics.*, **127**, pp. 453–472, 1983.

[59] Hanes, D. M. & Inman, D. L., Observations of rapidly flowing granular–fluid materials, *J. Fluid Mechanics*, **150**, pp. 357–380, 1985.

[60] Takahashi, T., Debris flow on prismatic open channel, *J. Hydraulic Div.*, ASCE, **106**, pp. 381–396, 1980.

[61] Chu, J., Basic characteristics of sediment–water mixture with hyperconcentration, *Proc. 2nd Int. Symp. On River Sedimentation,* Nanjing, China, pp. 265–273, 1983.

[62] Krieger, I. M. & Dougherty, T. J., A mechanism for non-Newtonian flow in suspensions of rigid spheres, *Trans. Society of Rheology*, **3**, pp. 137–152, 1959.

[63] Takahashi, T., Nakagawa, H. & Satofuka, Y., Newtonian fluid model for viscous debris-flow, *Debris-flow Hazards Mitigation: Mechanics, Prediction and Assessment*, Wieczorek & Naeser (eds), Balkema, Rotterdam, pp. 255–262, 2000.

[64] Major, J. J. & Pierson, T. C., Debris flow Rheology: experimental analysis of fine-grained slurries, *Water Resources Research*, **28(3)**, pp. 841–857, 1992.

[65] Julien, P. Y. & Lan, Y., Rheology of hyperconcentrations, *J. Hydraulic Eng.*, ASCE, **107**, pp.346–353, 1991.

[66] Lorenzini, G. & Mazza, N., From the phenomenology of debris flows to a rheological modelling approach validated with experimental data, 2003.

[67] Chen, L. C., General solution for viscoplastic debris flow, *J. Hydraulic Eng.*, **114**, pp. 259–282, 1988.

[68] Su, C. G. et al., Study on the velocity distribution of debris flow, *J. Chinese Soil and Water Conservation*, **24 (1)**, pp. 75–82, 1993.

[69] Takahashi, T., Satofuka, Y. & Chishiro, K., Dynamics of debris-flows in the inertial regime, *Proc. First International Conference Debris–flow Hazards Mitigation: Mechanics, Prediction and Assessment*, New York, ASCE, pp. 239–248, 1997.

[70] Franzi, L. & Bianco, G., The dispersive stresses range width as an indicator of debris-flows or bed load, *Debris-Flow Hazards Mitigation: Mechanics, Prediction and Assessment*, Wieczorek & Naeser (eds), Balkema, Rotterdam, pp. 351–359, 2000.

[71] Rickenmann, D. & Weber, D., Flow resistance of natural and experimental debris–flows in torrent channels, *Debris-flow Hazards Mitigation: Mechanics, Prediction and Assessment*, Wieczorek & Naeser (eds), Balkema, Rotterdam, pp. 245–253, 2000.

[72] Seminara, G. & Tubino, M., *Debris Flows: meccanica, controllo e previsione*, Presidenza del Consiglio dei Ministri, Dipartimento della Protezione Civile, pp. 284, 1993.

[73] Comini, G., *Lezioni di termodinamica applicata*, Padova, SGE, 1997.

[74] Jenkins, J. T. & Savage, S. B., A theory for rapid flow of identical smooth nearly elastic spherical particles, *J. Fluids Mechanics*, **130**, pp. 186–202, 1983.

[75] Egashira, S., Miyamoto, K. & Itoh, T., Constitutive equations of debris-flow and their applicability, *Proc. First International Conference Debris-flow Hazards Mitigation: Mechanics, Prediction and Assessment*, New York ASCE, pp. 340–349, 1997.

[76] Julien, P. Y., Dynamics of debris flow, *Recent Developments on Debris-Flows*, Aronne Armanini & Masanori Michiue (eds), Springer, pp. 65–69, 1997 .

[77] Cocchi, A., *Elementi di termofisica generale e applicata*, Esculapio, 1990.

[78] Liu, K. F. & Lai, K. W., Numerical simulation of two-dimensional debris-flow, *Debris-Flow Hazards Mitigation: Mechanics, Prediction and Assessment*, Wieczorek & Naeser (eds), Balkema, Rotterdam, pp. 531–535, 2000.

[79] Hashimoto, H. & Hirano, M., A flow model of hyper-concentrated sand–water mixtures, *Proc. First International Conference, Debris-flow Hazards Mitigation: Mechanics, Prediction and Assessment*, New York ASCE, pp. 464–473, 1997.

[80] Hungr, O., et al., Debris flow defences in British Columbia, *Rev. in Engineering Geology,* Geological Society of America, **7**, pp. 201–222, 1987.

[81] Okubo, S., Ikeya, H., Ishikawa, Y. & Yamada, T., Development of new methods for countermeasures against debris flows, *Recent Developments on Debris-Flows*, Aronne Armanini & Masanori Michiue (eds), Springer, pp. 166–185, 1997.

[82] Arattano, M., Recenti avanzamenti nelle misure di protezione di tipo passivo contro le colate detritiche torrentizie, *GEAM,* **3**, 1999.

[83] Julien, P.Y and Leon, C., Mudfloods, mudflows and debris-flows - classification, rheology and structural design, Invited paper at the *International Workshop on Mudflows and debris-flows,* Caracas, Venezuela, November 27–December 1, 2000.

Subject index

M

N

P

R

S

Glossary

A

Abrasion

Physical wearing and grinding of a surface through friction and impact by material carried in air, water, or ice.

Alluvial Fan

Large fan–shaped terrestrial deposit of alluvial sediment on which a braided stream flows over. Form as stream load is deposited because of a reduction in the velocity of streamflow.

Alluvium

Sediment that originates from a stream.

Angle of Repose

Measurement commonly used in civil engineering. It is the maximum angle at which a material can be inclined without failing. Geomorphologists use this measurement for determining the stability of slope to mass movements.

B

Basin

A topographic rock structure whose shape is concave downwards.

Bed

Sedimentary structure that usually represents a layer of deposited sediment.

Bed– Load

Portion of the stream load that is carried along the streambed without being permanently suspend in the flowing water.

Bedrock

Rock at or near (beneath soil and regolith) the Earth's surface that is solid and relatively unweathered.

Boulder

Large fragment of rock that has a diameter greater than 256 mm (200 mm in the United Kingdom).

C

Capillary Action
Movement of water along microscopic channels. This movement is the result of two forces: the adhesion and absorption of water to the walls of the channels; and cohesion of water molecules to each other.
Capillary Water
Water that moves horizontally and vertically in soils by the process of capillary action.
Clastic Sedimentary Rock
Sedimentary rocks that are formed by the lithification of weathered rock debris that has been physically transported and deposited by sedimentation processes. These rock fragments could vary in size from silt to large pebbles.
Clay
Grain fraction containing mineral earth particles with a size less than 0.004 mm in diameter. May include any kind of mineral material (mineral clays, silex, calcite, crystalline hydroxides and oxide minerals).
Creep
Slow mass movement of soil downslope. Occurs where the stresses on the slope material are too small to create a rapid failure. See soil creep.

D

Drainage
The ability of soil to eliminate the water it contains.
Debris
Any mixture of earth, rock, organic material and fluid incorporated within a debris flow.
Debris Flow
A type of mass movement where there is a downslope flow of a saturated mass of soil, sediment and rock debris.
Density (of Matter)
Refers to the quantity of mass per unit volume. For gases, density involves the number of atoms and molecules per unit volume.
Deposition
Laying down of sediment transported by wind, water, or ice.
Detrital Rock
Sedimentary rock that is composed of particles transported to their place of deposition by erosional processes. Examples of such rock include sandstone and shale.
Detritus
Shed tissues, dead bodies and waste products of organisms. In most ecosystems, detritus accumulates at the soil surface and other types of surface sediments.

E

Earthflow
A rapid type of downslope mass movement that involves soil and other loose sediments. Usually triggered by water saturation from rainfall.
Earthquake
It is a sudden motion or trembling in the Earth. The motion is caused by the quick release of slowly accumulated energy in the form of seismic waves. Most earthquakes are produced along faults, tectonic–plate boundaries, or along the mid–oceanic ridges.
Erosion
The removal of weathered sediment or rocks by the forces of wind, water and ice.

F

Flood
Inundation of a land surface that is not normally submerged by water from quick change in the level of a water body like a lake, stream or ocean.
Fluid
Substance, gas or liquid, that has the property of flow.
Fluid Drag
Reduction in the flow velocity of a fluid by the frictional effects of a surface.
Friction
Resistance between the contact surfaces of two bodies in motion.

G

Gravel
A term used to describe unconsolidated sediments composed of rock fragments. These rock fragments have a size that is greater than 2 mm.
Gravitational Water
Water that moves through soil due to gravitational forces. Soil water in excess of hygroscopic water and capillary water.
Gravity
Is the process where any body of mass found in the universe attracts other bodies with a force proportional to the product of their masses and inversely proportional to the distance that separates them. First proposed by Sir Issac Newton in 1686.
Groundwater
Water that occupies the pore spaces found in some types of bedrock.

H

Hazard
Phenomenon that can cause loss of life, injury, disease, economic loss, or environmental damage.
Hydrograph
A graph describing stream discharge over time.

I

Infiltration
The absorption and downward movement of water into the soil layer.

J

Joint
A fracture in a rock where no movement has taken place or where no movement has taken place perpendicular to the surface of the fracture. Important in rock weathering because it increases the exposed surface area.

L

Lahar
A very rapid type of downslope mass movement that involves mudflows from volcanic ash.
Landslide
Term used to describe the downslope movement of soil, rock and other weathered materials because of gravity.
Laminar Flow
Movement of water within a stream that occurs as uninterrupted parallel flows. Laminar flow generally occurs in areas where friction is low.
Liquefaction
Temporary transformation of a soil mass of soil or sediment into a fluid mass. Occurs when the cohesion of particles in the soil or sediment is lost. Often triggered by seismic waves from an earthquake. For this condition to take place the pore spaces between soil particles must be at or near saturation.
Liquid
A state of matter where molecules have the ability to flow and the surface of this mass displays the property of surface tension.
Lobe
A tongue–like extension of some material. For example, the ice lobe of an alpine glacier.

M

Mass
Refers to the amount of material found in an object (usually of unit volume).
Mass Balance
The relative balance between the input and output of material within a system.
Mass Movement
General term that describes the downslope movement of sediment, soil and rock material.
Mass Wasting
General term that describes the downslope movement of sediment, soil and rock material.

Matrix
The collective name for the sand, clay and silt components contained in soil, in all their possible combinations.
Mixture
We can distinguish between two types of mixtures. The former represents a physically homogeneous system (composed of a single phase) that is chemically heterogeneous (composed of a number of chemical types or components). The latter represents a physically heterogeneous system (composed of more than one phase) that can be either chemically hetero– or homogeneous (composed of one or more chemical type or component).
Movement
A term used in geography that deals with the migration, transport, communication and interaction of natural and human–made phenomena across the spatial dimension.
Mudflow
Form of mass movement where fine textured sediments and soil mix with water to create a liquid flow.

N

Natural Hazards
Natural phenomena that produce negative effects on life. The study of the hazards of natural phenomena.

P

Pebble
A rounded piece of rock that is larger than gravel.
Percolation
Vertical movement or infiltration of water from the Earth's surface to its subsurface. Movement usually stops when the flowing water reaches the water table.
Permeability
A measure of the ability of soil, sediments and rock to transport water horizontally and vertically. Permeability is dependent on the porosity of the medium the water is flowing through. Some rocks like granite have very poor permeability, while rocks like shale are actually quite pervious. As for soils, sand is the most pervious, while clay has the lowest permeability. Silt usually is somewhere in the middle.
Phase
Includes the various parts of a system in the same stage of aggregation.
Plastic Deformation
Irreversible change in the shape of a material without fracture as the result of the force of compression or expansion.

Porosity
The void spaces found in rock, sediment, or soil. Commonly measured as the percentage of void space observed in a given volume of substance. The vacuums (pores) are usually classified according to size.
Precipitation
Is any aqueous deposit, in liquid or solid form, that develops in a saturated atmosphere (relative humidity equals 100%) and falls to the ground generally from clouds. Most clouds, however, do not produce precipitation.
Pressure
Is defined as the force acting on a surface from another mass per unit area.
Prediction
Forecast or extrapolation of the future state of a system from current or past states.

R

Regolith
Loose layer of rocky material overlying bedrock.
Rock
A compact and consolidated mass of mineral matter. Three types of rock are recognized: igneous, sedimentary and metamorphic.
Rockfall
Type of mass movement that involves the detachment and movement of a small block of rock from a cliff face to its base. Normally occurs when the rock has well–defined bedding planes that are exaggerated by freeze–thaw action or thermal expansion and contraction.
Rock– Slide
Large–scale mass movement of rock materials downslope.
Runoff
The topographic flow of water from precipitation to stream channels located at lower elevations. Occurs when the infiltration capacity of an area's soil has been exceeded. It also refers to the water leaving an area of drainage. Also called overland flow.

S

Saltation
Transport of sediment initiated by moving air or water where particles move from a resting surface to the transport medium in quick continuous repeated cycles.
Sample
A sample is a subset group of data selected from a larger population group. Most samples are drawn at random to guarantee equal representation in the data.
Sand
Mineral fraction of the soil whose particles have a size between 0.06 and 2.0 mm in diameter.

Sediment
Solid material that has been or is being eroded, transported and deposited. Transport can be due to fluvial, marine, glacial or aeolian agents.
Seepage
The gradual movement of water into the soil layer.
Slow movement of subsurface water to the surface. This flow is not great enough to call it a spring.
Shear Rate
The velocity gradient measured across the diameter of a fluid–flow channel, be it a pipe, annulus or other shape. Shear rate is the rate of change of velocity at which one layer of fluid passes over an adjacent layer.
Shear Stress
Stress caused by forces operating parallel to each other but in opposite directions.
Silt
Mineral particle with a size between 0.004 and 0.06 mm in diameter. It is debris material from clastic rock, composed of various minerals such as quartz and clay material composed of angular granules with a compact structure.
Slope Failure
The downslope movement of soil and sediment by processes of mass movement.
Slump
Rotational slip. Form of mass movement where material moves suddenly along a curvilinear plane.
Snout
Front end of a glacier. Also called the terminus.
Soil
Layer of unconsolidated material found at the Earth's surface that has been influenced by the soil forming factors: climate, relief, parent material, time, and organisms. Soil normally consists of weathered mineral particles, including flood/river sands, gravel, clay, marine and lacustrine silt, colluvium and rock fragments, dead and living organic matter, air space, and the soil solution.
Soil Creep
Slow mass movement of soil downslope. Occurs where the stresses on the slope material are too small to create a rapid failure.
Soil Erosion
Transport of soil mineral particles and organic matter by wind, flowing water, or both. Human activities that disturb the soil surface or remove vegetation can enhance this natural process.
Soil Permeability
The rate at which water and air move vertically through a soil.
Soil Porosity
The volume of water that can be held in a soil. Also refers to the ratio of the volume of voids to the total volume of the soil.
Soil Water
The water found occupying the pore spaces between soil particles.

Solid

A state of matter where molecules where the mass of the substance does not have the property of flow.

Solid Concentration

The mean solid concentration by volume is defined as the ratio between the volume occupied by the solid fraction and the total mixture volume.

Solifluction

Form of mass movement in environments that experiences freeze–thaw action. It is characterized by the slow movement of soil material downslope and the formation of lobe–shaped features. Also see gelifluction.

Sorting

Process in which granules are selected according to their dimension, shape and specific weight under the action of transport agents and sedimentation mechanisms.

Stable Equilibrium

In a stable equilibrium the system displays tendencies to return to the same equilibrium after disturbance.

State of Matter

Form of matter. Matter can exist in three different forms gas, liquid and solid.

Static Equilibrium

Static equilibrium occurs where force and reaction are balanced and the properties of the system remain unchanged over time.

Stead–State Equilibrium

In this type of equilibrium the average condition of the system remains unchanged over time.

Stream

A long narrow channel of water that flows as a function of gravity and elevation across the Earth's surface. Many streams empty into lakes, seas or oceans.

Streambed

Bottom of the stream channel.

Stream Discharge

A river or stream's rate of flow over a particular period of time. Usually measured by a current meter and expressed in cubic mets per second. Stream discharge depends on the volume and velocity of the flow.

Streamflow

The flow of water in a river or stream channel.

Streamload

Refers to the material or sediment carried by a stream. It normally consists of three components: bed–load (pebbles and sand that move along the stream bed without being permanently suspend in the flowing water), suspended load (silts and clays in suspension) and dissolved load (material in solution).

Subsidence

Lowering or sinking of the Earth's surface.

Surface Creep

The sliding and rolling movement of soil particles on the Earth's surface because of wind. Eolian process of soil –particle movement.

Surface Tension
Tension of a liquid's surface. Due to the forces of attraction between molecules.
Suspended Load
Portion of the stream load that is carried almost permanently suspended in flowing water.
Suspension
Erosional movement of sediment continually held in the transport medium of air, water or ice.
System
A system is a set of interrelated components working together towards some kind of process.
System Boundary
Outer edge of system. Zone between one system and another system.

T

Talus
An accumulation of angular rock debris from rockfalls.
Talus Slope
A slope that is composed of talus.
Threshold
The level of magnitude of a system process at which sudden or rapid change occurs.
Traction
Erosional movement of particles by rolling, sliding and shuffling along the eroded surface. Occurs in all erosional mediums (air, water and ice).
Transport
One of three distinct processes involved in erosion. It is the movement of eroded material in the medium of air, water or ice.
Turbulent Flow
Movement of water within a stream that occurs as discrete eddies and vortices. Turbulent flow is caused by channel topography and friction.

U

Unstable Equilibrium
In an unstable equilibrium the system returns to a new equilibrium after disturbance.
Urbanization
Expansion of cities into rural regions because of population growth. In most cases, population growth is primarily due to the movement of rural–based people to urban areas. This is especially true in Less Developed Countries.

V

Vacuum
Space devoid of atoms or molecules.

Emptying of air.

Valley

A linear depression in the landscape that slopes down to a stream, lake or the ocean. Formed by water and/or ice erosion.

Velocity

The speed of movement of an object in one direction.

Viscosity

The amount of the resistance to flow in a fluid due to intermolecular friction.

Volume

The occupation of space in three dimensions. Measured in cubic units.

W

Wave

A moving swell or ridge on the surface of a solid or liquid or within the medium of a gas. Electromagnetic radiation also travels in waves.

Weather

The state of the atmosphere at a specific time and place.

Weathering

Physical, chemical or biological breakdown of rocks and minerals into smaller sized particles.

Wetting and Drying

Physical weathering process where rocks are mechanically disintegrated by the accumulation of successive layers of water molecules in between the mineral grains of a rock. Sometimes called slaking.

Printed in the United States
20838LVS00003B/211-225

9 781853 128028